On Complementarity

A Universal Organizing Principle

SERIES ON KNOTS AND EVERYTHING ISSN: 0219-9769

Editor-in-charge: Louis H. Kauffman *(Univ. of Illinois, Chicago)*

The Series on Knots and Everything: is a book series polarized around the theory of knots. Volume 1 in the series is Louis H Kauffman's Knots and Physics.

One purpose of this series is to continue the exploration of many of the themes indicated in Volume 1. These themes reach out beyond knot theory into physics, mathematics, logic, linguistics, philosophy, biology and practical experience. All of these outreaches have relations with knot theory when knot theory is regarded as a pivot or meeting place for apparently separate ideas. Knots act as such a pivotal place. We do not fully understand why this is so. The series represents stages in the exploration of this nexus.

Details of the titles in this series to date give a picture of the enterprise.

Published:

Vol. 69: *Mathematics of Harmony as a New Interdisciplinary Direction and "Golden" Paradigm of Modern Science*
Volume 3: The "Golden" Paradigm of Modern Science: Prerequisite for the "Golden" Revolution in Mathematics, Computer Science, and Theoretical Natural Sciences
by A. Stakhov

Vol. 68: *Mathematics of Harmony as a New Interdisciplinary Direction and "Golden" Paradigm of Modern Science*
Volume 2: Algorithmic Measurement Theory, Fibonacci and Golden Arithmetic's and Ternary Mirror-Symmetrical Arithmetic
by A. Stakhov

Vol. 67: On Complementarity: A Universal Organizing Principle
by J. S. Avrin

Vol. 66: *Invariants and Pictures: Low-dimensional Topology and Combinatorial Group Theory*
by V. O. Manturov, D. Fedoseev, S. Kim & I. Nikonov

Vol. 65: *Mathematics of Harmony as a New Interdisciplinary Direction and "Golden" Paradigm of Modern Science*
Volume 1: The Golden Section, Fibonacci Numbers, Pascal Triangle, and Platonic Solids
by A. Stakhov

Vol. 64: *Polynomial One-cocycles for Knots and Closed Braids*
by T. Fiedler

More information on this series can also be found at http://www.worldscientific.com/series/skae

Series on Knots and Everything — Vol. 67

On Complementarity

A Universal Organizing Principle

Jack Avrin

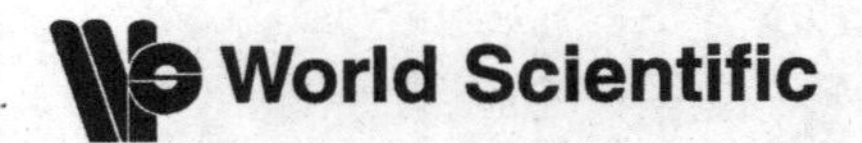

NEW JERSEY · LONDON · SINGAPORE · BEIJING · SHANGHAI · HONG KONG · TAIPEI · CHENNAI · TOKYO

Published by

World Scientific Publishing Co. Pte. Ltd.
5 Toh Tuck Link, Singapore 596224
USA office: 27 Warren Street, Suite 401-402, Hackensack, NJ 07601
UK office: 57 Shelton Street, Covent Garden, London WC2H 9HE

Library of Congress Cataloging-in-Publication Data
Names: Avrin, Jack, author.
Title: On complementarity : a universal organizing principle / Jack Avrin.
Description: New Jersey : World Scientific, [2021] | Series: Series on knots and everything, 0219-9769 ; Vol. 67 | Includes bibliographical references.
Identifiers: LCCN 2020040371 (print) | LCCN 2020040372 (ebook) | ISBN 9789813278974 (hardcover) | ISBN 9789813278981 (ebook) | ISBN 9789813278998 (ebook other)
Subjects: LCSH: Complementarity (Physics) | Matrices.
Classification: LCC QC174.17.C63 A97 2021 (print) | LCC QC174.17.C63 (ebook) | DDC 530.01--dc23
LC record available at https://lccn.loc.gov/2020040371
LC ebook record available at https://lccn.loc.gov/2020040372

British Library Cataloguing-in-Publication Data
A catalogue record for this book is available from the British Library.

For any available supplementary material, please visit
https://www.worldscientific.com/worldscibooks/10.1142/11234#t=suppl

Desk Editor: Soh Jing Wen

Typeset by Stallion Press
Email: enquiries@stallionpress.com

Printed in Singapore

I hereby dedicate this book to
Charlene Novak Avrin
With whom
I shared the joys and the sorrows of life itself

“*There are more things in heaven and earth, Horatio, than are dreamt of in your philosophy*”.
William Shakespeare, Hamlet, Act I, Scene 5

Preface

Not all books have a Preface; in many cases the Introduction suffice to expound upon the whys and wherefores of the book, enough to keep you reading it! However, now that you're already reading it, I must tell you that the preface to this book has a very personal significance to me for several reasons. One is that I feel duty-bound to write the book and the preface is a good place to explain why. The second reason — well, I just had another birthday (I'm 95! and it's unlikely that another preface to another book of mine will be rolling off the press any time soon). But the third reason is the one I really want to tell you about! I once received an accolade — that's right — in fact, a title!

So here's the story; it won't take long and there's sort of a moral to it. Many years (actually many decades!) ago I was admitted to the graduate physics department at the University of Southern California along with a job as a laboratory assistant for the undergraduate laboratories. $75 dollars a month plus free tuition courtesy of the G.I. Bill! And driving a war-surplus Jeep and living with my parents, what more could I want! The labs were also war-surplus, two story ex-army barracks with desk space for the Lab assistants. I soon realized that there was already a social structure headed by an acknowledged academic star, a somewhat older fellow — I'll call him Sam — with a penchant for characterizing his associates. It took a while but eventually I received my characterization; I was dubbed "The Master of the Obvious"!

The way it came to pass was that some of my fellow students were playing a lot of chess and they taught me how to play too. Well sort of; mainly it was so that they could practice their moves against a neophyte. Anyway, one day Sam (who was also the best player) came in with one of those chess problems — you know the kind that they have in the newspaper, or they used to. This one was "check and mate in one" and he predicted that the worst player would be the first to solve the problem. Sure enough I did, and it took me all of five or ten seconds to do so; it was obvious!

So what's the moral? I'll tell you that at the end of the preface; don't go away. In the meantime what I would like to tell you is what this book is about; something (often!) so *obvious* that when you tell people about it they recognize it immediately, "sure, it's obvious!" Yet at the same time it's so important, so *ubiquitous* and so far-reaching that it appears to constitute an *indispensable principle* for the existence of everything from life on this planet to perhaps the universe it inhabits. It's called the Principle of Complementarity just like it says on the cover of the book.

The concept of complementarity has a long history dating back to at least the ancient and honorable philosophy of Yin and Yang viewed in the Orient as a way of life (with a logo viewable as replicated on the cover of this book!). In the west it was resurrected by the Nobelist Niels Bohr who, in the mid 1920's, enunciated the Principle of Complementarity as the cornerstone of the so-called Copenhagen view of Quantum Mechanics (QM), something that came as a bit of a shock to both Werner Heisenberg whose Matrix theory emphasized the corpuscular nature of the elementary particles and Erwin Schrödinger who brought out their undulatory aspects in his famous wave equation.

In essence what Bohr maintained, was that the two views constitute an unavoidable yet complementary dichotomy with a resolution that depends upon the nature of what is asked of nature in any particular physical experiment. In their penetrating analysis of the epistemological quandaries of QM, George Greenstein and Arthur Zajonc (1997) devote an entire chapter to complementarity and the high drama of Bohr's presentation of it, as an overarching principle to the influential members of the physics community concerned with such matters, including Schrödinger, Heisenberg and Einstein. (He was able to convert only one of those three stalwarts.)

In any event, it turns out that Bohr's Principle and mine differ in some important aspects that originate in the microbiological world as well as in much of modern physics that Bohr could not have been aware of. So, you might consider this book as an update and amplification of Bohr's Principle.

And while I'm on the subject, there's something that's been bothering me for a long time and if you don't mind maybe I can take this opportunity to straighten it out; your forbearance gets you on the ground floor of those with knowledge on this important piece of cosmic insight. Actually, it's all very simple but it needs to be said. In the first place, Bohr developed his complementarity ideas at about the same time Heisenberg came up with his Uncertainty Principle. When young, unknown Werner, all aglow, rushed over to tell the world-renowned Niels all about it, instead of effusive praise, what he got (I'm guessing here, of course, based on what I've read here and there) was something like "that's nice; I'm glad you came up with another example of my most important Principle". Whereupon Werner's crest fell to the floor with a thud but, greatly offended, he began to argue the point. (Ineffectively, I might add).

Anyway, that's the picture I formed for myself! Of course Bohr was right; I hold Uncertainty in the highest regard, but it is really just an example, although a most important example of Complementarity. In any case, it does ***not*** merit the countless paragraphs expounding upon its' alleged mysteries to say nothing of its' philosophical implications. It's not mysterious at all; it's just a consequence, first, of the fact that Quantum Mechanics requires complex (both phase and amplitude) mathematics for its explication and second, that *the ubiquitous h*, Max Planck's quantum of action must be used as a normalization factor for each pair of variables that make up the uncertainty restriction. I'll say more about it in Chap. 11, "Quantum Mechanics and Radar"; it's really quite straightforward (which doesn't mean that it doesn't require some discussion!).

Well, so much for now about Complementarity and Quantum Mechanics. What got me involved with Complementarity in the first place was something completely different that emerged in the process of writing my previous book (Avrin 2015), published last year (see below), in which Complementarity emerges almost as an afterthought in the course of pursuing what seemed to be a resemblance between the algebraic formalism of DNA and that of my version of the elementary particles and which turned out to be most remarkable. Unfortunately, the importance and generality of that result percolated into my consciousness too late to do the spadework needed for its adequate inclusion in that book. I trust I've since done the job it deserves; let me know what you think.

As you undoubtedly know, the structure of the DNA molecule was discovered by James Watson and Francis Crick in 1953 (NLM 2021). In 1968, Watson wrote a book, "The Double Helix", in which he recounted his and Crick's discovery and their intense rivalry therein, with Nobelist Linus Pauling in the mid 1950's, to reach that goal first (Watson 1968). In the book, Watson credited Complementarity as the single feature of DNA that highlighted their victory. Here's what he said: "In almost any other situation Pauling would have fought for the good points of his idea. The overwhelming merits of the self-complementary DNA molecule made him effectively concede the race."

Overwhelming merits, indeed! Neither you nor I would be here to read and write this book, respectively, absent Watson's "merits"! And in this book, I shall endeavor to show how those same merits extend from DNA on to the fundamentals of physics and beyond. I shall have more to say about Complementarity below, but at this point I would like to enlarge upon another part of the title, namely *The Meaning of "Is"* and, for that, I need to tell you another couple of stories, both true.

Some of you may remember the first one: many years ago a well-known American politician — *really* well-known and occupying a high administrative position had been placed in an awkward defensive posture, because of some ill-advised personal proclivities and his responses to early queries thereto. In fact, he was being mercilessly grilled by a prosecutorial quorum of his peers. Being extremely intelligent and clever (although not always possessed of good judgment), he managed to evade disastrous admissions for a long time but, eventually, he found himself in a corner from which he could escape only by appealing to the ultimate logical response, namely "It depends on what you mean by ***is***". A loaded statement if there ever was one and very dependent on context!

In this book, I shall attempt to make some sense of *The Meaning of "Is"* in the context of some intricate and challenging matters of a philosophical as well as a physical nature. Which brings me to the second story: Recently, I acquired a book entitled "Our Mathematical Universe; My Quest for the Ultimate nature of Reality" (Tegmark 2014). It was written by Max Tegmark, a brilliant physicist of international repute who, in his writing, most certainly lives up to his standing in the scientific community. By coincidence I was given the book by an old friend, nominally in exchange for a copy of the book I had recently published (Avrin 2015). (Actually, it was published by the World Scientific publishing company as part of the series "on Knots and Everything" and I was just the author). I hope Professor Tegmark will forgive me for excerpting the main message of his book, namely that the "Ultimate Nature of Reality ***is*** Mathematics", a position he justifies by deep and cogent argumentation.

Of course I have a motive in underscoring and highlighting "is". The motive is that in my book, in fact in the preface, I emphasized my ***explicit*** belief that the UN of R is ***not*** mathematics! So you can imagine my consternation — "OMG! What do I do now" or words to that effect; there I was, on record as being directly contradictory to a world-famous scientist on a subject that both of us had stressed! Well, the way the matter came up in my book is like this: I had been writing about a couple of books I read a long, long time ago, one being Gulliver's Travels, nominally about coping with existence in various extreme societies and people. For example, by extremely large individuals or extremely small ones, but actually an ***allegorical*** treatment of the way things were in the England of the Author's day. I used that to discuss the nature of both the universe and our microscopic makeup as follows:

"In a way, what I've been describing here is a parallel to Gulliver; the (immensely difficult to contemplate) universe we inhabit on the one hand and the (realistically impossible to picture) microworld of which we are composed on the other. So, what I'm trying to say is that the best we can do with our theories, our models, our grand summarizations or our detailed itemizations of either the large or the small, is to use what amounts to *metaphor* to propound *allegorical* patterns of what we perceive, structures we can understand, that appear to be self-consistent (according to how we define such criteria) and that we can use to make predictions (or post-dictions). We can do this in words, or in pictures, but inevitably in mathematical relationships that some say are all we can ever hope to put together, in fact that constitute the ultimate reality. Well, I think, maybe not quite; I would say, instead, that the mathematics *constitutes an indispensable metaphorical tool, a language that, in translation, also provides* an *allegorical, albeit a more formal, portrayal.*" (I had not yet learned of Tegmark's book).

The main emphasis of that book, is what I have called "An Alternative Model of the Elementary Particles of Physics", an alternative that is to the currently iconic Standard Model, and much of the detailed description of my model derives from material published over the period 2005 to 2012, although I really started thinking seriously about the whole subject back in the fall of 1996. Nevertheless, it's only gradually that I began to feel I could discuss it in a more comprehensive way and it took a while before I finally realized that there's a rather elementary topic that needs to be clarified, to wit, "just exactly what ***is*** an elementary particle?" Or, for that matter, what we mean by "elementary"? Or, even "particle". To say nothing of what we ***mean*** by "is". And I included the following:

The celebrated nobelist, **Eugene Wigner**, who did so much to advance the role of symmetry in physics is credited with a detailed *algebraic* analysis of the subject with the conclusion that an "elementary particle 'is' an irreducible unitary representation of the group, G of physics, that is, the double (universal) cover of the Poincare group of those transformations of special relativity which can be continuously deformed to the identity". Whereupon my commentary thereto was:

"Most impressive, but not exactly what we'd like to see here, in a book of this nature; although proven useful to some of the scholars among us, it provides scant guidance for describing, say, the nature of the particulate occupation of space, in some ontologically satisfying way — somehow, a rather unsettling state of affairs." The professor had a bit more to say and so did I, but my conclusion was. "... I don't think one should say that "an elementary particle ***is*** an irreducible, etc." but, to adhere strictly to my thesis, something like "can be viewed for analytical purposes" would be more like it.

So, as you see, I've been preoccupied with *The Meaning of "Is"* for some time. I'm going to talk about it some more in the Introduction and other places but, since Complementarity is the main topic in the book, let's go back and dwell on it for a while. First, as per the definition, I hereby stipulate that this book is mainly about situations where two or more "entities" — not necessarily "things" (see below) — acting ***together*** are ***necessary*** to make some other entities happen or be important; the necessity is important here.

Examples abound, some simple, some quite complex, some every-day and mundane but some so involved and arcane you might need a lot of background to really understand them. Complementarity is indeed ubiquitous. Not only that, it's all over the place and you can't avoid it although I don't know why you'd want to. For one thing, you wouldn't be here without it because, believe it or not, complementarity makes our lives possible! Which, you'll agree is rather important. I'll talk about that in more detail in the rest of the book but it's a demonstrable fact.

And, I should add, complementarity also makes life worth living! Which, of course, is a matter of opinion but, as far as I'm concerned, is just as true; I wouldn't have it any other way. Some of life's greatest pleasures are complementarily enjoyed. I'm sure you can think of lots of them so I need not enumerate them here, but I'd like to mention one I consider an amazing demonstration of complementarity in action. Have you ever listened to a virtuoso pianist play a work of consummate beauty, combined with the boundless creativity and complexity instilled in it by the genius of a major composer? The listener can be entranced by the tonality and the patterns and rhythms of its progression and spellbound by the player's dexterity all at the same time. It's as if he or she were really two people, yet it's done with only the two hands of a single person, each playing a completely different score, both *written* by a single person so as to act together in a manifestation of his musical vision! So, we have the complementarity of the composer and the pianist, the two parts of the score and the pianist's two hands. If that's not complementarity at its peak of perfection I don't know what is!

Tonality, rhythm, patterns, counterpoint, variation, repetition, etc., — they're not confined of course to the piano; they are the stock in trade of ensembles of any type and size — duets, string quartets, rock groups, orchestras, choral groups — you name it, and their practitioners learn how to complement each other to best effect. I can go on and on, citing for instance the complementarity demanded of the players of team sports. Or teams of participants in any project of consequence (which, unfortunately includes warfare) for that matter. Or an economic system large or small that relies on the division of labor between individuals with different skills and responsibilities. Or simply conjugal relationships! I say simply but historically that's been a bedrock of societies throughout the history of the human race. Actually we see it in some species throughout the animal kingdom as well and it's a beautiful thing to behold.

Even simple, everyday human relationships and interactions employ complementarity — it's inherent really. We meet and shake hands: complementarity. Carry on a conversation; complementarity (although not always well balanced!). Every group, every family learns the drill, the balance of activity and attitudes that works for it. If they don't, it doesn't work and the group can't last. Or do you play tennis? If so, you generally expect the ball to come back to you after you hit it, right? Complementarity!

And how about physiology? Here we're mostly talking about individuals. Well, for bipeds or quadrupeds there's basic locomotion like walking or running or even swimming; those limbs have got to act complementarily, or we stumble; or drown (sorry about that). Then there's dancing; as they say: "It takes two to Tango" and they had better know how to complement each other. And speaking of hands, talk to anyone who has lost the use of one of them, even temporarily and they'll tell you how hard it is to wash one hand without the complementarity of the other.

I should note that complementarity manifests itself as either physical, i.e. tactile, visible, etc. or mental — an idea, concept or, indeed, a ***principle***! (which, of course was my hesitation to use the word "thing" above). Or, in more philosophical terms, seen from either the Ontological or Epistemological points of view. Even the purely mental, e.g. the mathematical, philosophical or spiritual spheres are not immune from complementarity. Indeed, it may be said that the first examples of the principles of complementarity in human activity are documented in the ancient philosophy of the orient as exemplified in what came to be called, the philosophy of Yin and Yang. But aspects of complementarity continue, of course, to influence human activity and relationships. For example, the simple act of communication, so crucially human in its sophistication, is an exercise in complementarity. Right now, you're reading something I wrote: it would have been pointless to write it absent the expectation of a readership, hopefully one encompassing a broad background of criticality.

Similarly, the union of a speaker, actor, musician etc., and an audience even, in modern times, one that receives the message via recording, is a complementary phenomenon that provides the originator of the information with valuable guidance. This is *feedback*, also a common phenomenon with examples in an immense range of disparate fields, from physiological processes to control systems to economic systems, are examples that all involve a *forward* sector that fashions and transmits a "command" to a variable system, device, etc. and a *backward* sector that transmits information regarding the actual condition or activity of the system. Comparison of the desired and actual behavior is then used to fashion the "command" and ideally their difference disappears. The two sectors may be viewed as operating in a complementary fashion to implement a desired activity or regulatory function.

Well, so much for a broad-brush treatment of complementarity. There's probably a lot more we ought to say about it as a *principle* but I've probably already been talking too much so let's proceed with the book.

Oh, by the way, before we do that; about that chess problem: I don't recall any discussion about it at the time but the way I can reconstruct it is that there was somewhat of a scarcity of pieces on the board as a consequence of which the more experienced players made an assumption that inhibited them from guessing the winning move. Given my inexperience, I did not make that assumption and was not so inhibited. That's all I'm going to say about it at this time; if you're experienced you can probably figure out what transpired and if you're not and still want to know, send me an email and I'll reply with the answer as I recall it! (I still don't know much about chess!).

So what's the moral to the story; here it is: "one must be wary of making unwarranted assumptions!" The reason those other guys didn't solve the puzzle right away is that they made an unwarranted assumption based on their experience. And if you think about it, much if not all of the progress in physics, and in fact in science altogether is based on realizing and discarding the presence of unwarranted assumptions that had evolved over time to be the accepted wisdom. I'll try not to do that here, so, with that in mind, let us proceed.

As we do, you will notice, I'm sure, the ***ubiquitous*** appearance of the following neat little 2 × 2 Matrix (actually, you can't avoid it because it's at the heart of the book and, I believe, of what we perceive as ***reality!***)

$$m = \begin{pmatrix} 1 & -1 \\ -1 & 1 \end{pmatrix}$$

with the eigenvectors

$$\begin{pmatrix} 1 \\ -1 \end{pmatrix} \quad \text{and} \quad \begin{pmatrix} -1 \\ 1 \end{pmatrix}.$$

Early on, you can expect to see it in terms of the way DNA does its thing to create us creatures but we meet up with it throughout the manuscript, sometimes with the help of ***four legendary*** particles known as ***Pions*** (see below) that are expected ***to make things happen*** in the domain of the elementary particles — and so they do!

Now, each of the four Pions is recognized by its ***own spelling*** (simply by two letters of the English alphabet). However, when we

evaluate the actual "***physical makeup***" of ***each letter of each of the four*** (which I did) we find that the ***2 × 2 matrix of Pions*** is ***identical to our matrix***

$$m = \begin{pmatrix} 1 & -1 \\ -1 & 1 \end{pmatrix}!!$$

Amazing! Wow and Holy Cow! And Save me a Suite in Sweden! Especially, when we consider the low-key history of how the ***Pions*** were discussed in the first place by ***Enrico Fermi*** and **C.-N. Yang (Fermi and Yang 1949)**. And, on the other hand, how we find ***Matrix m*** to be introduced in its ***own*** special (Algebraic) way in ***Sections IV and V of this book.*** So; the wondrous point here is that the necessary work is done, as required, by one or more of the four workers, but identified ***either*** as one or more ***Pions or as Matrix m depending on the situation!***

So: it's beginning to look like ***we have already discovered*** a prime example of ***Complementarity*** and we haven't yet got to the main part of the book! (Maybe I should start looking for that Suite in Sweden after all!)

Well, that's all I have to say in the Preface; there's more detail about Pions in the book and quite a lot about Matrix m. Also there's a short Appendix presented mainly to show how ***Pions*** facilitate the ***decay*** of the Neutron but please don't just skip to the end of the book; there's a lot of really interesting stuff in between (Starting with the Introduction!) that covers a lot of variety. The Table of Contents, coming up, indicates that!

Contents

I
Preliminaries

1

Introduction: Complementarity as a Principle

The ruminations in the preface show just a smattering of the applicability of complementarity. Clearly the subject is immense: the title of this monograph alone would indicate that! So, I hope you're not too disappointed when I tell you that this book is not primarily about the things I mentioned above. Rather, it emphasizes something of a completely different nature but every bit as important if you believe the nature of the physical world is important. It's mostly about fundamental physics with a very important early foray into elementary biochemistry, DNA, no less. Then it reintroduces what I have called an Alternative Model of the Elementary Particles, alternative, that is to the well-known, even iconic Standard model. That model was originally developed in a series of papers published in the journal JKTR (Avrin 2005, 2008, 2011, 2012a), then, as more or less collected in the journal Symmetry (Avrin 2012b), and finally, in the book alluded to above (Avrin 2015).

Eventually the current book generalizes, encompassing much, I hope, of the world of Physics that you and I might agree is important, (depending upon background and imagination) not everything I'm sure, but a lot and I hope you enjoy it. I hope you even find it credible: I certainly did! But, before I go on, I need to ask you to pause and consider with me: Is it not true that we spend our lives from the moment we're born trying to make sense of the world around us, to impose some kind of order on what we perceive? To discern patterns — of behavior, human behavior, or more generally animate behavior, or more generally yet, of the physical world we encounter. We need patterns to predict what's likely to occur from one moment to the next, from one episode to the next, long range or short, so we can live our lives or plan projects, avoid danger, satisfy our needs, chart the course of nations, etc.

Musicians are concerned with the progressive patterns of tonality, rhythm or theme; Writers with those of words, sentences, ideas and the effective use of language. Mathematicians talk of symmetries and the group structure thereof. In ordinary human affairs when the pattern is broken, sometimes we think of it as perverse. Perhaps you've heard the phrase "the unreasonable perversity of inanimate objects" if it occurs in the case of entities over which we have little control. On the other hand, in fundamental Physics, as some of you may know, pattern ***breaking*** is an integral part of fundamental theory but in the case of the written word or in music it calls for rewrite because the written word is subject to the perversities of Language and in music it might just not sound right!

Mathematics is basically also a language or, an analogy with those spoken in our world, a set of languages, Algebra, Geometry, even Algebraic Geometry, Differential Calculus to give us rates of change, Integral Calculus to give us manifolds of various order, and Differential Geometry to operate in detail in or on them, Euclidean for space that's "flat" and Riemannian if it's not. Plus, Mathematical logic to assess the nature of them all. More varieties of mathematics are being developed all the time but the original impetus, I believe was, and still is to have a way to express spoken or written language about the physical world, its attributes and their relationships, more "precisely", more cogently, more economically and more formally constrained to obey rules of grammatical order and logic. Pattern breaking is not supposed to occur, and it doesn't if the rules are followed. Nevertheless, mathematics has inherent limitations.

There was a time when eminent mathematicians believed that one could start in some fundamental field with a set of axioms that would permit proofs of all subsequently obtainable results. Their eminences were riding a veritable Tsunami of mathematical advances so you can't really fault them for what turned out be a vainglorious attempt at logical self-consistency in their field of endeavor. The individual who upset the self-consistency applecart was Kurt Gödel, a little known logician of either Czech, Austrian or German citizenry, depending on the outcome of the nationalistic stirrings of the period (as well as his preference!). In 1931, Gödel demonstrated in his celebrated "Incompleteness" Theorem that, given a particular mathematical field, there exist mathematical statements known to be irrefutably true but whose veracity or falsity cannot be determined within the body of the chosen field, that is on the basis of the selected axioms — they are "***undecidable***" (Raatikainen 2021).

Many years ago (It must have been around 65 years ago) I was a graduate student in Physics but I enrolled in a class on Mathematical Logic given by a Professor Henkin, an acknowledged expert in the field. There was no text but Henkin led us through Gödel's demonstration in detail as the course progressed, so we diligently filled many pages with course notes up to the final "aha!" moment at which point Henkin paused and simply stood for a long moment, expectantly facing the students who stared blankly back. At which point, yours truly blurted out in my inimitable fashion: "But that's just (somebody's — right now I can't remember whose) ***paradox***!" Which got me an instantaneous, automatic A in the course! I should confess that I had taken a course on logic (not mathematical logic!) the previous semester where I learned about paradoxes. Maybe that should be ***paradoxi***; it's from the Greek and is defined in various ways in the Dictionary.

Here's a couple:

1. A statement that is self-contradictory in fact and, hence false.
2. A statement that seems contradictory, unbelievable or absurd but may actually be true.

The situation I like is where, for example, there's a statement on each side of a piece of paper regarding the veracity of the statement on the other side. In essence, there is an inherent paradox when the two statements are mutually contradictory thus leading a reader to consider the message on one side and then the other, back and forth, in an oscillatory manner without being able to arrive at a decision as to which side if either is believable; you've most likely encountered it. Both sides are of course necessary to the paradox and are therefore

complementary and it turns out that the ***oscillatory*** behavior associated with it is exactly what we find early on in this book as the mathematical ***signature*** (see below) of Complementarity.

Oscillatory behavior may also be seen as paradoxical in the sense that, for example, the angular orientation of a pendulum is out of phase with its time rate of change so that when the bob is at its maximum orientation, it is when there is a maximum force to reverse direction, a situation that is also analyzed in the book. This is of course only one example out of many, some of which are also discussed in the book, but I mention it because it has bearing on the modeling of elementary particles as presented early on. I am also reminded of my dim past in the Defense/Aerospace industry and the preoccupation with stability in the design of feedback control systems which are liable to break into incontrollable oscillations especially if there exist unanticipated feedback loops.

In a way, paradoxes are scary stuff; for example, if you're a mathematician and you end up with a paradox in some attempt at proving something, you know you're stuck — nowhere to go. On the other hand, it gives you an important message: there is something illogical or illegitimate in what you were trying to do in the first place. Which, of course, is what it told those who were bent on establishing the self-consistency of mathematics. But you may have begun to wonder where this discussion is going. Well, I think *The Meaning of "Is"* ***is*** also fraught with unanticipated paradoxical behavior. That simple-sounding phrase harbors a world of hidden "meaning". For one thing, we're up against the inherent difficulty of language to express basic ideas or feelings in absolute terms unaccompanied by circular reasoning that can result in paradoxical situations.

When we say something ***Is***, what exactly do we mean? Do we mean we can see it, hear it, touch it? Predict how it will change? Understand its relationship with other notions? That's generally alright if we're talking about physical entities in the realm of ordinary experience but, as we know, we have to be careful, especially down in the quantum world where we end up with the constraints of Heisenberg's uncertainty principle. And of course, right now in this paragraph, I'm trying to formulate a way to describe *The Meaning of*

"*Is*" in the more general sense in which it can refer to intangibles — ideas, statements, principles and I'm about to run into a circularity problem. It would appear that *The Meaning of "Is"* is a barrier, a polysemic thicket that bars our way to ultimate understanding.

And when we say that mathematics ***is*** the Ultimate Reality we're additionally stuck, not only because of *The Meaning of "Is"* but the meaning of "Reality" as well and, most certainly, the meaning of Ultimate Reality — Ultimate? Stuck and stuck again. How are we, the limited creatures that we are, struggling to impose order on the ***ordinary*** reality we are capable of perceiving, going to recognize "Ultimate Reality" when (or if) we're faced with it? Do we have anything to compare it to? And, perhaps most pertinently, the use of a particular piece of mathematics is not necessarily confined to a particular subject or phenomenon. In fact, if you recall, the purpose of this book, namely, to document the universal ubiquity of complementarity, emerges in the previous book as a result of the commonality we find in the algebraic structures of DNA and the set of elementary particles of my Alternative Model.

In any event, right now, I'm skeptical. In fact, for all practical purposes, I seriously doubt that such a "thing" as "Ultimate Reality" exists, at least in a form that creatures like us can aspire to understand. And I don't feel especially isolated in harboring that belief. In 2007, John D. Barrow[1], a distinguished cosmologist and professor of mathematics as well as a prolific author of penetrating books on various aspects of the subject, published his ***second*** book on the theories of everything, this one being entitled "New Theories of Everything" (Barrow 2007), the purpose being to take into account the advances in a variety of sciences pertinent to the subject that had occurred in the interim. One wonders when (or if) the good Professor would feel it advisable to write a ***third*** book on the subject! Nevertheless, I would not hesitate to refer anyone who would dispute my point of view to him.

All things considered I'll settle for Complementarity but as a ***Principle*** as per the title of the book. It's not the "Ultimate"

[1]Whom you'll meet later on as the coauthor of another book.

anything; nor the "Theory of Everything". It's just not that kind of entity. However, I do think that it constitutes a universal ***necessity***, the requirement that must be obeyed by the laws of Physics and, indeed, for existence in the physical world. One reason I think so is its elegance — the elegance of simplicity. As we shall see in the early chapters of this book, the symmetry patterns with which Complementarity can be expressed — its group structure — is the ultimate (!) in simplicity.

Here, I think it's pertinent to quote Professor Barrow as follows. "...If we are to arrive at full understanding of complex systems, especially those that result from the haphazard workings of natural selection, then we shall need more than current candidates for the title 'Theory of Everything (TOE)' have to offer. We need to discover if there are general ***principles*** that govern the development of complexity in general, which can be applied to a variety of different situations without becoming embroiled in their peculiarities." (*My emphasis*).

I think the Principle of Complementarity (POC) qualifies as one such for a very large variety of situations even though, as you will see, this book features considerable "embroilment" in their various "peculiarities" on the premise that such embroilment bolsters confidence in the conclusions! I think the problem with the TOE is encapsulated in the phrase — perhaps you've heard it delivered in rather sonorous tones — "What Can Man Know". Sonorous or not, the message therein is really fundamental to who and what we are and our place in the universe. In any event, assuming that the POC is not the TOE, what can we say about their relationship? What I can say — and it's really the thesis of this monograph — is that when (and if) there is something reliably (!) identifiable in some way, shape or form as a TOE, it *must incorporate* complementarity as an essential attribute.

Some people interpret the TOE as the reconciliation of Quantum Mechanics and the Theory of General Relativity, otherwise known as the Theory of Gravity. That's it. Subject closed. I disagree. At least, I would put it in a different way. At minimum, I should think, there ought to display or perhaps at least derive the underlying nature of Space and Time (or better Spacetime), and of all the elementary particles it supports plus their interrelationships and interactions; in other words, fundamental physics. So, the main thing this monograph will emphasize is how pervasive ***complementarity*** is in fundamental physics. However, even as constrained to the nature of the physical world, complementarity is not limited to physics, fundamental or not; in fact, it is ubiquitous over an immense span of scale, activity and classification. Also, a variety of mathematics, especially geometry. In terms of the physical sciences, in addition to physics it encompasses biology, physiology, and the study of all manner of life forms that inhabit our planet and their interrelationships.

Well, that's almost enough rumination for now. I still have some notions about the relationship between Complementarity and *The Meaning of "Is"* that I would like to share for your review and commentary but for now, I would like to show you how I got started with complementarity in the first place and how as I widened my outlook, the certainty emerged from writing ***this*** book is something I really had to do! I'll begin with some elementary biochemistry all wrapped up in a geometrical package. The subject is Deoxyribonucleic Acid (DNA) — the double Helix, but just a barebones description to begin with, followed in Chapter 3, by a stripped-down description of the Alternative Model of the elementary particles of physics as per the previous book with a few asides about complementarity as it arises. Next, we generate, in Chapter 4, a detailed ***comparison*** of the ***structure*** of DNA and that of our elementary particles, going on to display a simple, ***algebraic*** expression that characterizes the complementarity of ***both*** DNA, and the ***elementary particles*** of physics as depicted in the Alternative Model, something that actually emerged in the previous book but which we make a bit more explicit herein. Finally, Chapter 5 demonstrates the explicit significance of that expression to both subjects, especially to the action of DNA in ***self-replication***. As the book progresses we expand the domain of applicability so that, finally, the use of the term ***Universal*** is, if not completely justified, at least justifiable.

2

Deoxyribonucleic Acid, The Molecular Ladder to Life on Earth

The two main physical scales of interest here are that of the elementary particles as exemplified by the Alternative Model and that of microbiology as exemplified by DNA. At this point I must stipulate that in addition to being neither a Physicist nor a Mathematician, I am also not a Microbiologist. Nevertheless, there's no harm in me repeating what is quite generally known, namely the famous double helix which consists of two potentially parallel strands winding about each other. When (or if) they are uncoiled, the two strands taken together would exhibit a ladder-like structure encompassing, as rungs, organic molecules known as "nucleotides" of which there are four varieties, or more accurately, two sets of two. Each member of one set is composed of ***three*** molecules, a phosphate, a sugar and a base known as a ***Pyrimidine***, which has a hexagonal ring-like structure with a particular atom at each of the corners. Each member of the other set has the same phosphate and sugar and a base known as a ***Purine*** composed of ***two*** ring-like structures, one hexagonal and one pentagonal, linked together.

The phosphate and sugar in each of the four are also linked together to form one segment in the resulting sequence of alternating phosphates and sugars that constitutes ***each leg*** of the ladder. The "***rungs***" of the ladder are of course where the familiar "Genetic Code" resides and it works like this: First of all, the detailed ***chemistries*** of the two Pyrimidines (their atomic constituents and the locations thereof) differ and so do the detailed chemistries of the Purines, thus giving us the familiar Pyrimidines, ***Thymine*** and ***Cytosine*** (T and C) and the two Purines, ***Adenine*** and ***Guanine***

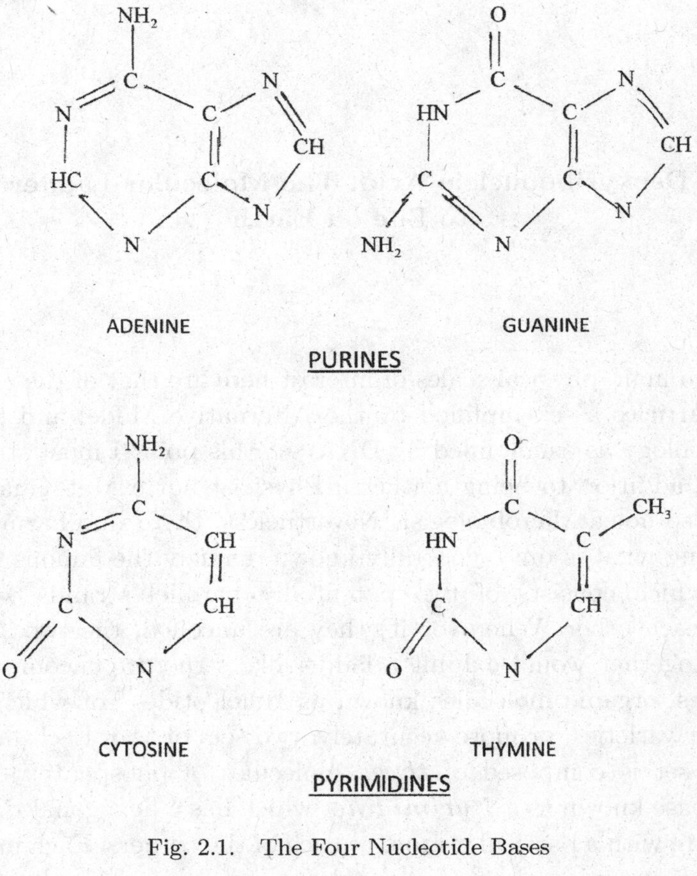

ADENINE GUANINE

PURINES

CYTOSINE THYMINE

PYRIMIDINES

Fig. 2.1. The Four Nucleotide Bases

(A and G), a ***fourfold*** set (and, as we shall see, the first obvious, top-level connection to the basic four of our Alternative Model). Fig. 2.1 shows the four bases, the Phosphate-Sugar segments of the legs of the ladder, having been omitted as of little concern in what follows.

At this point, detailed geometry enters into the picture: each of the four bases is securely linked to a sugar molecule of a segment on one leg or another of the ladder, T and C by a nitrogen atom of their hexagonal ring and A and G similarly by such an atom of their pentagonal ring, in such a way as to extend ***inward*** relative to the ladder and normal to its legs. If we concentrate on either leg, we see what might appear to be a random sequence of letters taken from the four-letter alphabet (A, G, T and C), but we know it's not random; it's how a genetic code is spelled out. And although there is of course an element of randomness in the creation of a DNA molecule in the first place, it occurs in quite a different way that we need not pursue herein.

As it is also well known, what Watson and Crick did was to construct, without regard to the genetic code, a large, laboratory 3-D mock-up scale of the molecule's structure on the basis of all kinds of inputs including X-ray photos of DNA samples and chemical experiments conducted in laboratories in a number of countries (NLM 2021). We pick up the story at a point where most of the mistakes and false leads were over and the pair had more or less settled on a two-strand helix, as per the above, on the outside of some kind of two-by two pairing of the four nucleotides for the rungs. By that time a lot of detailed chemical information was available to the DNA community including some inexplicable statistics on the relative amounts of the four nucleotides; there always appeared to be ***similar amounts*** of Adenine and Thymine on the one hand and of Guanine and Cytosine on the other.

Finally, we know that each "***rung***" consists of a Purine-Pyrimidine combination, each member of the pair being securely connected to one leg or the other on its "outside" but the pair is only lightly bonded together in the interior of the rung by what is known as a Hydrogen bond. The reason for this pairing — and this is crucial — is that Watson and Crick were trying for a smooth structure and, because of their relative sizes, pairing of Purines together and Pyrimidines together made for awkward bulges and twists; the former were too long and the latter too short. To make a too-long story shorter, the choice of pairing G with C and A with T not only made a ***perfect fit*** but explained the relative abundances found in chemical experiments.

Fig. 2.2 shows a highly simplified, short segment schematic of DNA. This example illustrates perfectly what Watson was talking about as quoted in our Preface ("the overwhelming merit of the self-complementary DNA molecule") the ***complementarity*** of G and C

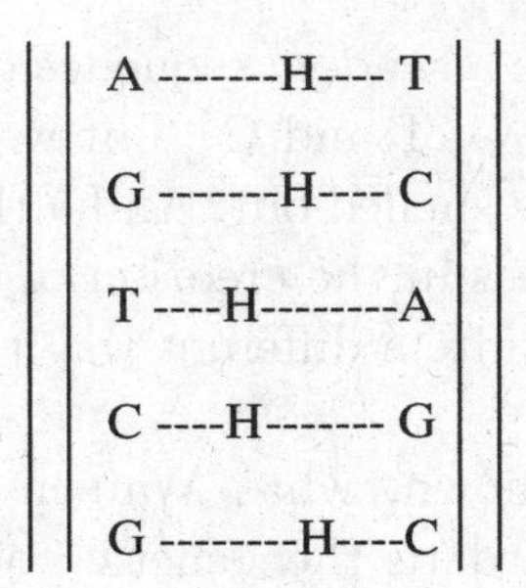

Fig. 2.2. Simplified Segment of DNA

and of A and T such that if and when the ladder structure "peels apart" (those hydrogen bonds break easily) we end up with two ***complementary sequences*** of nucleotides, each member of which is just waiting for another complementary partner with which to mate (very suggestive depending on how your brain works). And of course in the nucleotide-rich sea interior to a cell, that's exactly what happens and we end up with two complete, identical DNA molecules where there was only one before! Thus begins the growth of a complete, brand-new member of the species!

For me, the preceding describes a phenomenon so amazingly ***elegant*** in its fundamental simplicity and yet so fraught with the potential for boundless complexity, the complexity of life on Earth — it boggles the mind! Even though it immediately evokes fundamental questions of belief, we know it has populated the earth with countless species that perpetuate themselves but are still capable of evolving. You will recall once more how Watson alluded to the "overwhelming merit" of the complementarity that distinguishes the double helix model. In hindsight, I would phrase the matter in somewhat stronger terms: DNA works ***because of its complementarity***! In fact, as far as we are concerned, it is the most dramatic manifestation of the ***Principle of Complementarity***; one can state unequivocally: ***we are who we are***, individually and collectively, as a direct result of that manifestation!

In essence, what we are saying here is that the ***Concept of*** DNA, the way it works, is superlatively ***elegant*** because it is really very

simple, in fact about as simple as it can be and still generate the amazing variety of species we encounter on this planet. We might say that DNA manifests an ***economy of concept*** even though, to the best of our knowledge, it was not actually conceived but rather emerged as a process of evolution. Which is about as far in this subject I had better proceed at this point.

However, I think we can also say something similar about the Principle of Complementarity in itself; as a principle, it too manifests an ***Economy of concept*** as we shall repeatedly demonstrate in the rest of this book.

3

Alternative Model Taxonomy

Little in the preceding says much about the purpose of DNA which, of course, is to implement the genetic makeup of a new individual, something accomplished by means of the information lodged in the genes, each gene being a particular organization of a particular, nominally-sized number of what we have characterized in the preceding as "rungs" of the DNA ladder.

What we are interested in ***here*** is really something we can compare to the taxonomy of the Alternative Model of the Elementary Particles in terms of DNA. It's really just what we saw back in Fig. 2.1 which, in essence, is just a 2×2 matrix of molecules — basic molecules, of course!

So, what we need now is a comparable set of basic elementary particles and for that we need to do a little borrowing from my Book of Reference (BoR) (as little as possible!). However, I like to take advantage of authoritative encouragement any chance I get and so, first here's a natural: in 1949 two of physics most iconic figures, **Enrico Fermi** and **C. N. Yang** wrote a paper entitled "Are Mesons Elementary Particles?". The introduction to the paper contains the following statement (Fermi and Yang 1949):

"We propose to discuss the hypothesis that the *π-meson* may not be elementary, but may be a *composite particle* formed by the association of a *nucleon and an anti-nucleon.* The first assumption will be, therefore, that both *an anti-proton and an anti-neutron exist*, having the same relationship to the proton and the neutron, as the electron to the positron. Although this is an assumption that goes beyond what is known experimentally, we do not view it as a very revolutionary one. We must assume, further, that between a nucleon and an anti-nucleon, strong *attractive forces* exist, capable of *binding* the two particles together." (Emphasis added).

As things turned out, the "binding" mentioned in the quotation is visually ***manifested*** in the Alternative Model featured in the previous book and, in fact, it initiates and continues to implement the *expansion* of the model's ***taxonomy***, from a small set of basic elementary particles (in fact Fermions) to an edifice we can compare to the (also currently iconic) Standard Model of the elementary particles. Although that expansion is modeled in terms of ***algebraic geometry***, a salient feature of the model is the description of the "elementary particles" as "localized distortions in-and-of Space and Time, or better (since Einstein and Minkowski) in-and-of spacetime. In fact, the elementary particles exist as ***Solitons***, continuously, seamlessly forming and reforming the stuff of spacetime even as they move along within it."

Specifically, our particles begin life in Spacetime as ***Sine-Gordon*** solitons, actually in the form of ***Moebius Strips*** (MS) that can be viewed as ***concatenations of Torus Knots***, each taken from a small set of such namely the set of $(2, n)$ torus knots, meaning n revolutions in ***latitude*** for every two in ***longitude*** of the implicit toroidal "last". Figure 3.1 shows a typical concatenation of a small set for $n = 3$, the Trefoil knot. Some important results of the ***differential geometric*** analysis conducted in the previous book are based on this kind of portrayal and we shall delve into that further on in this book.

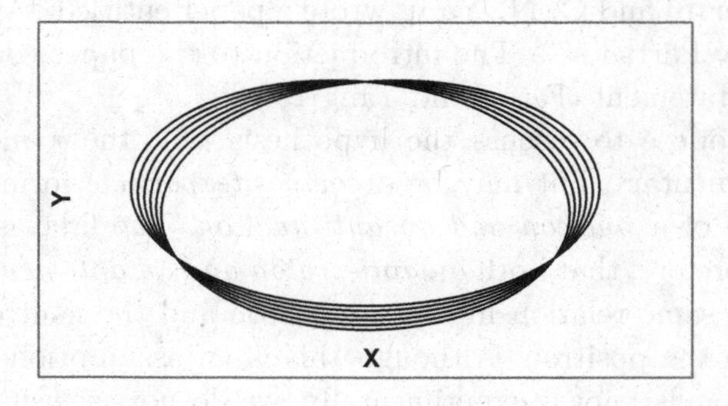

Fig. 3.1. Concatenated Trefoil Knot

Nevertheless, also as per the previous book, the ***algebraic*** geometric description of the model actually begins with the ***flattening*** of each such solitonic entity into a triangular configuration in $(2+1)$ spacetime. (Time is to be considered as measured normal to the plane of each figure) and the choice of a direction for the flow of the explicit solitonic distortion as shown in Fig. 3.2.

We see four individual, Flattened Moebius Strips, hereinafter designated as ***FMS*** (Imagine a very large number of the concatenations described above for each!), with both left and right twist for both $n = 1$ and 3 and with an implied "***direction of traverse***". These four constitute the basic ***Fermions*** of the model. The triangular planform is the ***minimal*** such configuration that can accommodate the twists of a Moebius strip and it is readily demonstrated that there can be exactly ***four*** such, no more, no less, two with one half-twist (both left and right) and, similarly, two with three half-twists. The corners of each figures are called "***Quirks***" (Not Quarks!) and are a ***binary*** set relative to the ***direction*** of traverse in that they bend either ***down into or out of*** the page.

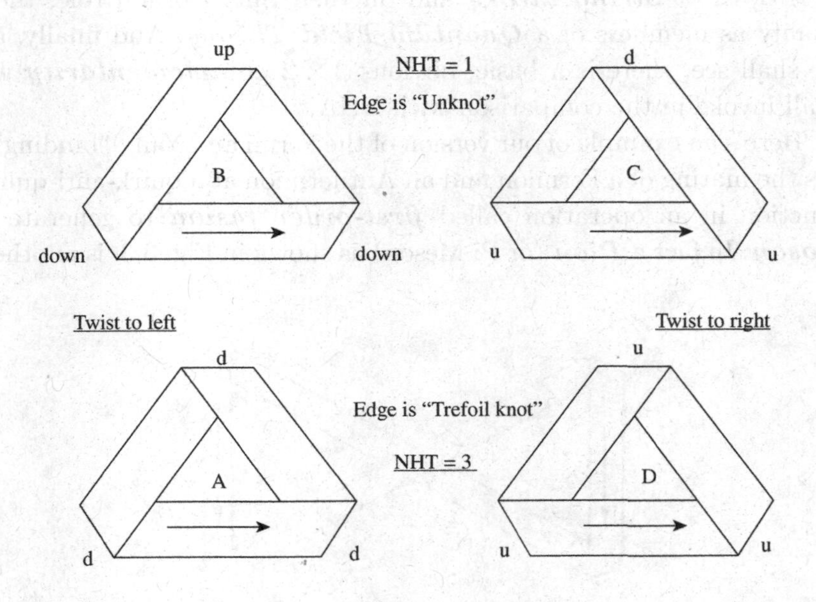

Fig. 3.2. The Four basic Fermions

Also associated with each such quirk, by definition, is an electric charge of either $+2/3$ or $-1/3$ of the charge of Proton for "up" quirks or "down" quirks, respectively. (There is a similar set of ***antifermions*** which differ from the first set in that their direction of traverse is ***reversed***.) As a result, the total charge associated with the set of four basic fermions is -1, 0, $+1$ and $+3$, respectively, which plots (versus twist) as a straight line with an average charge value of $+1/2$. We shall have more to say on this subject in what follows, but the main point of interest here is that twist and charge emerge as ***complementary*** concepts by virtue of the basic particle structure.

We really ought to interject at this point that these four FMS constitute a perfect example of ***complementarity*** at a very basic level; although they are indeed to be thought of as ***particles***, so defined as to function as such, at the same time, we recognize them as ***solitons***. They are also, by definition, ***waves*** which, as expressed above, "form and reform of the stuff of spacetime even as they move along within it"! We also note that, as ***Moebius strips***, they find residence, as the simplest ***Calabi-Yau*** manifolds, in M Theory, the outgrowth of ***String theory*** and, in their Sine Gordon role, they qualify as members of a ***Quantum Field Theory***. And finally, as we shall see, there is a basic, obvious 2×2 ***complementarity*** we shall invoke in the comparison with DNA.

Here's an example of our version of the Fermi and Yang "binding"; it's the mating of a Fermion and an Antifermion at a quirk-anti-quirk junction in an operation called ***first-order fusion*** to generate a ***Boson***. In fact a ***Pion*** (or Pi Meson) as shown in Fig. 3.3, is another

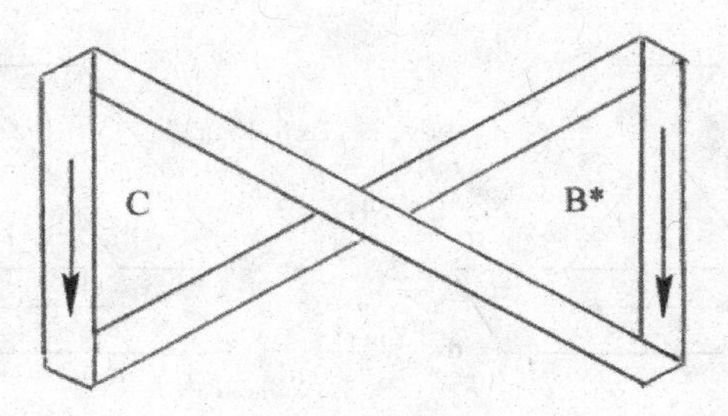

Fig. 3.3. Making a Boson by Fusion

example of ***Complementarity*** and, as will be shown, related to some very basic physics.

In other words, associating fermion C with the ***proton*** and B with the ***neutron***, (the asterisk indicates the antiparticle) represents a ***manifestation*** of exactly what Fermi and Yang were talking about, in this case the ***positively charged pion***! Note that it has zero twist because it joins basic FMS with twists of ± 1. There is only one restriction but a most important one: in order to maintain traverse, we can only join a ***fermion*** to an ***antifermion***, in fact, a ***quirk*** to its conjugate ***antiquirk***. In this case, the junction in the figure unites an "up" quirk, u, with its conjugate, u^*.

Appending another Fermion is correspondingly called ***second-order fusion*** and results in the AM version of higher order hadrons including the excited states of the basic four described above. One of the benefits of this approach is that the ***complementarity*** that characterizes the basic set of fermions extends to the entire taxonomy. Another is that the MS paràmeter of ***twist*** emerges as a manifestation of the particle quantum parameter of ***Isospin*** as well as a means for organizing electric charge subgroups, just one of a sizeable list of results that emerge sometimes seemingly unbidden.

From a more formal point of view, we considered an abstract, *group theoretic* approach that bypasses the detail but summarizes the general architecture of the taxonomy. In this (top-level) approach a particle hierarchy is developed as the *direct product of* ***vector spin spaces*** *parameterized by* ***spin***. Correspondingly, the abstract result is expressed as the direct sum of subsidiary spin spaces, the so-called Clebsch-Gordan decomposition. With the additional recognition that the applicable group structure is that of the ***gauge group*** *SU(2)*, the result of the direct product of vector spin spaces with spins S1 and S2 (a ***complementary*** procedure!) is expressible as

$$V_{\mathrm{S1}} \otimes V_{\mathrm{S2}} \rightarrow V_{|\mathrm{S1}-\mathrm{S2}|} + V_{|\mathrm{S1}-\mathrm{S2}|+1} + \cdots + V_{\mathrm{S1}+\mathrm{S2}} \qquad (3\text{-}1)$$

which equals $V_0 + V_1$ (particle spin = 0 or 1) for the case of first-order fusion, that is, for

$$\mathrm{S1} = \mathrm{S2} = 1/2.$$

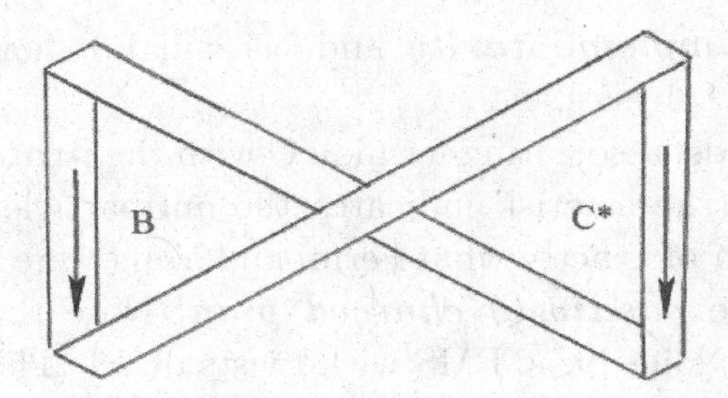

Fig. 3.4. The Boson Conjugate to Fig. 3.3

To begin with, we are specifically concerned with the *entire* ***vector*** *of four* basic spin 1/2 *fermions* $V = (A, B, C, D)^{\mathrm{T}}$ and its conjugate vector of *antifermions*, $V^* = (A^*, B^*, C^*, D^*)^{\mathrm{T}}$ and the direct product

$$M = V \oplus V^{*\mathrm{T}} \tag{3-2}$$

the result being the (self-adjoint) matrix of sixteen two-element composites shown in Eq. (3-3) where, in analogy with quantum mechanics, we note that M can be viewed as an ***operator***

$$M = \begin{bmatrix} [AD^*] & BD^* & CD^* & DD^* \\ AC^* & BC^* & CC^* & DC^* \\ AB^* & BB^* & CB^* & DB^* \\ AA^* & BA^* & CA^* & [DA^*] \end{bmatrix} \tag{3-3}$$

Note the two sets of three fermions each, (A, B, C and D, C, B), in the grouping of six matrix elements in the lower left-hand corner and the similar grouping of another six elements in the upper right-hand corner. As we shall see below, this is the first glimmering of the need to divide our taxonomy into two ***complementary***, ***overlapping*** sets (with BC* and CB* in common), a schism with far-reaching consequences. Notice also the two elements in *square* brackets in two corners of the matrix; in reality, these elements ***cannot exist*** because) no common quirk-antiquirk combination exists in either case.

Also, there are actually *only* ***two*** V_0 elements, namely, CB*, the one shown above in (3-3), above, and its conjugate, BC*, shown in Fig. 3.4, above.

Although, as per the above, these two elements can be formed by fusion, topologically, they are both just *doubled-over versions* of the trivial, zero-twist MS. In other words they are ***excited states*** of the basic ***untwisted*** state and, in fact (we recall), in each case the algebraic sum of the ***twists*** of the fused constituents is zero. The other twelve bosons are all V_1 vector bosons in their ground state and can also be formed either by fusion or directly by a twist whose Number of Half Twist (NHT) is also the sum of those of its constituents.

We must emphasize that the four bosons in the middle of the grouping of (3-3) are what Mesons Fermi and Yang were talking about, the four "Workhorses" that implement the Second order fusions that create Fermions with two basic fermions and an antifermion (See SectionV).

4

The AM/DNA Comparison

As mentioned in the Preface, a fundamental mathematical resemblance between DNA and our Alternative Model (AM) was established in my Book of Reference (BoR) (Avrin 2015). Here, we want to show this in a more detailed way. Nevertheless, although there is a lot of ***Complementarity*** associated with the AM, for comparison to DNA, all we really need are the basic set of ***fermions***, and it is clear in the preceding chapter that the development of the AM taxonomy is essentially just an enlarged reflection of the essence of that set. So, here's what we know about them:

1. In the first place, there are four of them.
2. Each one has three features, the quirks.
3. The quirks are labeled using a binary alphabet.

Similarly, the basic ***information*** associated with DNA resides in the ***nucleotides*** and there are also four of them as well; in fact that was one of the main reasons to wonder about a possible resemblance between the two structures in the first place. However, at this point that would appear to be all we have — the ***number four***! Nevertheless, maybe we can make that number a little more succinct. So, suppose we ***relabel*** the two varieties of "rung" (AT) and (GC), on the two-strand DNA ladder as, respectively,

$$\mathrm{AT} \Leftrightarrow \mathrm{R_A} \text{ and } \mathrm{GC} \Leftrightarrow \mathrm{R_G}.$$

Further, note that each of these rung varieties can occur in a directional way (in the ***direction*** normal to the legs), say from "left" to "right" — call that (+); and, conversely, from "right" to "left" — call that (–), the implication being that we need a more detailed labeling as, for instance, both

$$\mathrm{R}_{\mathrm{A}(+)} \quad \text{and} \quad \mathrm{R}_{\mathrm{A}(-)}.$$

For instance, we see both AT and TA, respectively in Fig. 2.2 and, similarly, both

$$\mathrm{R}_{\mathrm{G}(+)} \quad \text{and} \quad \mathrm{R}_{\mathrm{G}(-)}.$$

Again, a fourfold set but expressed in a kind of non-specific way as composed of two ***varieties*** each with two algebraic signs. But if we express the four in a two-dimensional array as

$$\begin{pmatrix} \mathrm{R}_{\mathrm{G}(-)} & \mathrm{R}_{\mathrm{G}(+)} \\ \mathrm{R}_{\mathrm{A}(-)} & \mathrm{R}_{\mathrm{A}(+)} \end{pmatrix},$$

it looks just like our fourfold basic FMS array discussed in Chapter 3

$$\begin{pmatrix} B & C \\ A & D \end{pmatrix}.$$

If, that is, the latter is also re-expressed, but here, in terms of both magnitude and sign of ***twist***, for instance as

$$\begin{pmatrix} T(-) & T(+) \\ T^3(-) & T^3(+) \end{pmatrix},$$

with "T" for twist. In other words, the upper row represents B and C, each with ***one*** half twist (N=1) to the left and right respectively, and similarly, the lower row represents A and D, each with ***three*** half twists, also to the left and right. The implication is that our basic set of FMS and the set of basic DNA nucleotides have at least the same top-level ***structural*** representation, still a top-level commonality but we no longer have to cite just a shared number four!

However, we don't have to leave it at that: consider the detailed makeup of the quirks for each particle in the set of four basic fermions of the AM. They can be summarized by this array:

$$\begin{pmatrix} B & C \\ A & D \end{pmatrix} \Rightarrow \begin{pmatrix} \mathrm{ddu} & \mathrm{duu} \\ \mathrm{ddd} & \mathrm{uuu} \end{pmatrix}, \tag{4-1}$$

quite succinct enough for our purpose here but we can go further: recalling how, in Section II, the notion was introduced that our basic fermions exist as 2 + 1 dimensional entities, suppose, additionally that the quirks can be viewed as very small steps in time, then we can identify an up quirk and a down quirk with time as follows:

$$d \Rightarrow i\tau \quad \text{and} \quad u \Rightarrow -i\tau$$

where

$$i = \sqrt{-1},$$
$$\tau = c\delta t,$$

δt is the size of the step in time and c is the speed of light. This is in a conceptual spacetime where time (but not space) is considered to be an imaginary dimension. Furthermore, suppose we view the quirks as ***operators*** so that the arithmetic here is ***multiplication*** (rather than the addition which, you may recall, we used to identify particle A with the electron, B with the neutron or a neutrino, and C with the Proton).

The new representation of our basic four fermions as a 2D array then becomes

$$\begin{pmatrix} B & C \\ A & D \end{pmatrix} \Rightarrow \begin{pmatrix} \text{ddu} & \text{duu} \\ \text{ddd} & \text{uuu} \end{pmatrix} \Rightarrow \begin{pmatrix} 1 & -1 \\ -1 & 1 \end{pmatrix} \tau^3 i \tag{4-2}$$

(you do the math!), which, represents a small, alternating group with only ***two*** elements namely $\pm i$, thus bringing the representation of our basic set down to bare essentials as an elementary expression of ***complementarity***! Which is as far as we can go; you can't get much simpler than back and forth — or up and down, etc. — unless you want to stick with only one or the other, i.e. do nothing, neither of which expresses complementarity.

Actually for our purposes we don't really need the explicit time representation because all the pertinent algebraic information is conveyed by the matrix on the right-hand side of Eq. (4-2) which, if we set $u = 1$ and $d = -1$ gives us a matrix

$$m = \begin{pmatrix} 1 & -1 \\ -1 & 1 \end{pmatrix} \tag{4-3}$$

for which we find the eigenvalues to be 0 and 2 (see next chapter) and the eigenvectors are

$$\begin{pmatrix} 1 \\ -1 \end{pmatrix} \text{ and } \begin{pmatrix} -1 \\ 1 \end{pmatrix}.$$

The question then is can we do something similar with the nucleotides? On the face of it, it certainly seems doubtful. In a way, the problem is an embarrassment of the riches, a rather shopworn phrase but apt here. Consider again Fig. 2.1 of Chapter 2 showing the basic chemistry of Purines and Pyrimidines; you have to admit that those structures are somewhat more complex than that of the four basic elementary particles! Furthermore, they express a completely different language. So how can we bring that kind of representation down to something that's at least reminiscent of the elementary particles?

Well, in that case, the elementary geometry of our particles really comes down to the presence of only ***three*** "***features***", namely the three ***quirks***, each with a ***label*** from an alphabet of just ***two*** letters. So, here's the plan: to begin with, we need to reduce the number of features per nucleotide diagram and one way to do that is to subdivide each diagram into ***regions***, each of which will include a number of atomic constituents, and that will be ***labeled*** such that each diagram is represented by a ***set*** of labels. The idea is to compare the four sets of labels and note the ***pattern*** of coincident and non-coincident labeling that results.

The trick is to do the subdividing in a judicious manner and, to begin with, it was noted that the pentagonal loops looked like an obvious object of special treatment since they are associated only with the Purines. That being the case, the final pattern of subdivision that emerged after a bit of cogitation is shown in Fig. 4.1 and the associated pattern of coincidences, indicated by a (+), and

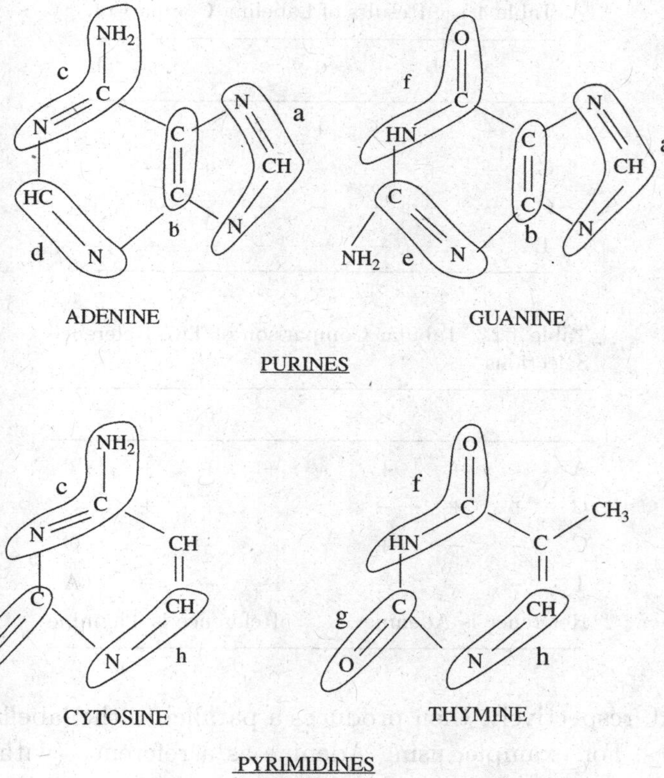

Fig. 4.1. Nucleotide Bases with Subdivisions

non-coincidences indicated by a (−), is shown in Table 4.1. The table was constructed with Adenine as a reference so that all four of its regions were denoted by (+).

As we see, Adenine has four labeled areas with the labels a, b, c and d, not all of which apply to all of the four nucleotides. However, the diversity of chemical structure (Of course necessary for the nucleotides to do their job!) has resulted in the need for more labels, a total of eight: a, b, c, d, e, f, g and h. Nevertheless, a closer look at the table reveals that labels a, b and c suffice (d, e, f, g and h are superfluous) to do the job; that is, equating + and − to

Table 4.1. Results of Labeling Comparison

	a	b	c	d	e	f	g	h
A	+	+	+	+	−	−	−	−
G	+	+	−	−	+	+	−	−
C	−	−	+	−	−	−	+	+
T	−	−	−	−	−	+	+	+

Table 4.2. Tabular Comparison of Two Reference Selections

	a	b	c	f	g	h	
A	+	+	+	+	+	+	T
G	+	+	−	−	+	+	C
C	−	−	+	+	−	−	G
T	−	−	−	−	−	−	A
Reference is Adenine				Reference is Thymine			

u and d, respectively, then produces a parallel to the labeling of our particles. For example, using Adenine as a reference (with labeling +++) produces the sequence A, G, C and T, which, as we see in Table 4.2, corresponds to the sequence D, C, B and A of our particle model. In other words, we have shown at least one way to validate the resemblance between our particle model and DNA not just superficially but on a ***detailed level***!

Upon closer scrutiny of Fig. 4.1, however, we see another way to achieve the same result: use ***Thymine*** as a reference. In this case, labels f, g and h suffice and a, b, c, d and e are superfluous.

These two tables are left-to-right mirror images, but they are ***not complementary***; to demonstrate that we need to show the ***broken*** symmetry in evidence from the point of view of either of the nucleotides as reference, and seen from the associated helical strand after separation as discussed above. We show that in Table 4.3 for all four nucleotide combinations using Adenine as a reference, but it becomes apparent if any of the other three are used in that way.

Table 4.3. Complementary Nucleotide Tables, (Reference is Adenine)

	a	b	c	c	b	a	
A	+	+	+	−	−	−	T
G	+	+	−	+	−	−	C
C	−	−	+	−	+	+	G
T	−	−	−	+	+	+	A

The table on the right is drawn simply by rotating the table on the left 180 degrees about an axis normal to the paper just to make the complementarity obvious at a glance but a closer look shows that the sequence abc $= + + -$ for Guanine is matched by its complementary sequence abc $= - - +$ for Cytosine and so forth for the other three nucleotides A, C and T and their complements T, G and A. Furthermore, it is straightforward to show the same relationships obtain using T, G or C as references. But, notice: no matter what reference is used, the same ***complementary*** pairings obtain: A with T and G with C! Which, as Crick and Watson noted, explains the statistical findings they had puzzled over. It looks like we're getting somewhere!

So now: back to the question posed above as to whether we can reduce the representation down any further, as we did for the particles. Well, suppose we start with the representation $\begin{pmatrix} G & C \\ A & T \end{pmatrix}$ for the nucleotides with, as in the above, Thymine as the reference. (Again, we can use any of the other three nucleotides as the reference). Then, using Table 4.2, this translates to $\begin{pmatrix} +-- & -++ \\ --- & +++ \end{pmatrix}$ which is essentially the quirk representation for the particles, except for the latter's use of imaginaries. (Again, we can use any of the other three nucleotides as the reference). Again we treat the symbols multiplicatively to arrive at $\begin{pmatrix} + & - \\ - & + \end{pmatrix}$, again, a two-state, algebraic representation of ***complementarity*** as for the particle case.

We can make the identification precise if we ***quantify*** this matrix as $m = \begin{pmatrix} 1 & -1 \\ -1 & +1 \end{pmatrix}$ at which point it would appear that we have ***ratified the assertion*** that DNA and our Alternative Model (AM) of the Elementary Particles share the same ***algebraic structure***, an exceedingly, one might say, surpassingly, simple in form ***signature*** for the ***complementarity*** exhibited by both DNA and the AM set of four basic Fermions.

Finally, still in terms of comparison, one thing we haven't talked about much is the requirement for symmetry breaking. Well, in the case of ***genetics***, it's the need to have both a Purine and a Pyrimidine base, with their different ***lengths***, (lightly) bonded together so as to make up each "codon" of the Genetic Code in order that we end up with two ***complementary*** codes after the "unzipping" and, in the end of the process, two copies of the original DNA molecule. In the case of the elementary ***particles*** of the AM it is due to a necessary choice for a ***direction*** of traverse of each FMS and we'll talk about that later.

5

The Signature of Complementarity; Replication Unleashed

The previous chapter brought out a gratifyingly elegant result, namely that we can write a simple algebraic entity, the matrix m, that, in essence can be viewed as a common ***signature*** for ***both*** DNA and the set of four ***particles*** that are basic to the taxonomy of the Alternative Model (AM) of the elementary particles, a result actually documented in the previous book. However, the "gratification" does not end with just the ***statement*** of the matrix; there's more and to explore it, we begin by finding the associated eigenvalues. As usual, an eigenvalue equation is found by setting the associated determinant to zero, thus:

$$\|m - \lambda I\| = \left\| \begin{matrix} (1-y) & -1 \\ -1 & (1-y) \end{matrix} \right\| = 0 \tag{5-1}$$

which produces

$$\lambda^2 - 2\lambda = 0, \tag{5-2}$$

the eigenvalues and $\lambda = 2, 0$ and (here's where the gratification kicks in!) the fact that matrix m ***satisfies its own eigenvalue equation***! That is we can set $\lambda = m$ to get

$$m^2 = 2m = \begin{bmatrix} 2 & -2 \\ -2 & 2 \end{bmatrix} \tag{5-3}$$

which is to say that there exists a successor to matrix, m, namely

$$m_2 = \begin{pmatrix} 2 & -2 \\ -2 & 2 \end{pmatrix} \tag{5-4}$$

whose eigenvalues we find to be $\lambda_2 = 4, 0$ *and the* ***new eigenvalue equation***

$$\lambda_2^2 - 4\lambda = 0 \tag{5-5}$$

which is satisfied by

$$\boldsymbol{m}_2^2 = 4\boldsymbol{m} \tag{5-6}$$

(note that m itself is still the original matrix) whose associated ***successor matrix*** is

$$m_3 = \begin{pmatrix} 4 & -4 \\ -4 & 4 \end{pmatrix}. \tag{5-7}$$

So that finally, as we ***continue*** the procedure we produce the ***binary*** series,

$$m_n = 2^{n-1} m, \tag{5-8}$$

which in turn, evaluates to $m_n = m, 2m, 4m, 8m$, etc. for $n = 1, 2, 3$, etc., thus ***mirroring the process of DNA replication*** in the nucleus of the cells of a ***living creature*** to make more of the same!

In summary to this point, not only can matrix m be viewed as the algebraic ***signature*** of ***complementarity***, but, similarly, the binary series

$$m_n = 2^{n-1} m \tag{5-9}$$

can be viewed as the ***signature equation*** of the seemingly endless biochemical ***replication of the DNA*** molecule[1] until — voila —, ***another member*** of the species appears! In essence, that sequence may, ***in turn***, be viewed as the output of a ***particular*** kind of ***unstable feedback process*** that ***apparently*** requires a ***unique*** matrix, the ***Complementarity Signature matrix***, first, to ***begin*** the process of DNA ***replication***, and then to be ***fed back*** upon each iteration!

---

[1]With humble apologies to James Watson and Francis Crick.

At this point, however, we must be careful what we say and how we say it because, as pointed out to me by Professor ***Joel D. Avrin***, University of North Carolina (UNC), Charlotte, while it is certainly legitimate to characterize our complementarity **signature matrix** m as begetting the ***particular***, unstable binary feedback process that leads to ***DNA replication*** as per the above, there exist ***other matrices*** that can result in the identical binary ***mathematical*** sequence. In other words, as far as non-specific, binary sequences are concerned, matrix m is sufficient but not necessary; that, to begin with, several other matrices ***equivalent*** to m, up to a coordinate ***transformation***, exist that can also produce that sequence. These are

$$\begin{bmatrix} 2 & 0 \\ 0 & 0 \end{bmatrix}, \begin{bmatrix} 0 & 0 \\ 0 & 2 \end{bmatrix}, \text{ and } \begin{bmatrix} 2 & 0 \\ 0 & 2 \end{bmatrix},$$

and ***any matrix*** in the binary ***sequence*** may also be viewed as the progenitor of those that follow.

Nevertheless, it is only matrix m which ***satisfies its own eigenvalue equation*** that is readily shown to emerge from the nature of ***both*** DNA and our fundamental ***Fermions*** and to reveal itself as leading to the binary sequence of ***DNA replication*** and ***that is matrix*** m, which is therefore responsible for ***species survival!!***

Finally, another point of interest: it turns out that there is also a straightforward, (and not surprising) connection between all *four* matrices discussed above and the justifying famous Pauli matrices, something that will be demonstrated in Chap. 24.

Some Discussion: We note that, as per the above, the general resemblance of the signature matrices for our Alternative Model (AM) particles on the one hand and DNA structure on the other was made precise by ***quantifying*** the latter's algebraic representation, a transformation that preserves its structure. In justification for this step we note that the above allows us to demonstrate an ***algebraic*** version of the way in which DNA replicates itself. Such ***mathematical*** quantification is ***validated*** by the resulting emergence of the simple algebraic ***representation*** of the actual physical replication process that takes place to create ***each individual of a given species***! But conversely, the physical ***process*** may be said to constitute a particular ***manifestation*** of the ***algebra***!

In other words, ***living species*** as we know them on this "Tiny blue planet" may be said to depend for their survival on a ***fundamentally unstable process***, the replication of DNA; something to think about!

In contrast, quantification of the corresponding Alternative Particle Model's signature matrix (which emerges simply by virtue of the previously defined values for the ***quirks***) needs no such explicit validation. In fact, we recall that, rather than replicating the process carried out above, the AM converts the original 2×2 matrix of basic Fermions (which we shall hereinafter refer to as the Quartet!) into a ***4-vector*** which, when multiplied by its conjugate, creates a 4×4 matrix of the products of first-order fusion, thus beginning a process that validates its existence as a ***manifestation*** of the gauge group $SU(2)$ and the further emergence of a valid ***taxonomy*** and comparison with the Standard Model. Not to mention a most gratifying list of automatically-emerging connections to well-known basic requirements and other models and systems of elementary physics.

Of course, we must emphasize that, as noted in the previous book, the algebra associated with the ***gauge group*** does not actually ***dictate*** the actual ***physical*** process, which is modeled as "first and second order fusion, beginning with the basic "quartet" but we are gratified to see that it does, in fact ***validate*** it. Nevertheless, in any case, we cannot confuse the algebra with the actual process. For example, in the case of DNA, the process of replication is complicated at every step by the ***requirement*** for an almost endless supply of the four basic nucleotides that must be available in order to complete each newly separated half of the DNA molecule. But these are readily available in the nucleus of every cell!

However, what about the Alternative Model? Here, the equivalent of the set of four nucleotides is the 2×2 "Quartet" we identified above and the goal is the construction of a ***taxonomy***. So the first question here is then how many such Quartets are needed just to activate the process and where can we get them? Well, as based on the 4×4 matrix that results from multiplying the associate 4 vector by its complex conjugate, (the matrix that begins the process of taxonomy building) we need ***four of them*** because each member of the vector must fuse with ***four*** members of the conjugate vector (except, of course, for the two corners of the 4×4 matrix where fusing is impossible).

And that's just the start; the next phase of the process is to construct a ***three-dimensional array*** housing all the three element terms resulting from ***second-order*** fusion. Thus, without going into detail at this point (in this book) it is clear that many more Quartets are required. And that's just for ***one realization*** of a taxonomy.

So, in the absence of Quartet-filled nuclei as in DNA replication, one can justifiably ask where did all the Quartets come from that we need to satisfy the demands of ***matter*** in the Universe? It would appear that we require an inexhaustible supply of them which implies that there can be only one answer to that question; the Universe at that stage in its development must be awash with quartets of four elementary Fermions and their conjugates! Which, at this point in the book, looks like a segue into ***Cosmology*** is in order.

Well, we will have a bit more to say about the apparent implications later on but, as a practical matter, an adequate discussion must await publication of the second edition of the previously referenced book but for now, we need to proceed with our widespread discussion of ***complementarity***. Before we leave the current subject, however, there are a couple of matters I'd like to bring to your attention:

First, from an epistemological point of view, it is possible to discern an underlying existential equivalence between our signature matrix and the ancient ***Yin-Yang*** symbol of the Orient, but, as mentioned in the preface, with a slight difference; where the traditional Yin-Yang complementarity is a ***duality***, our signature version may be characterized as a ***quaternary***! Following up on that notion,

Fig. 5.1. Yin-Yang Equivalent of Signature Matrix

consider the two-dimensional rendering of a signature "type" matrix shown in Fig. 5.1 with elements consisting of ***two*** such symbols (as seen either ***horizontally or vertically***) that together comprise ***four***, rather than just two elements colored either black or white. Then each such element can be viewed as forming a Yin-Yang symbol with another element on either its left or its right (or above or below it).

Finally, one more comment about DNA: Many years ago, the world famous, Nobel prize winning physicist Professor ***Erwin Schrödinger*** wrote a book entitled "***What is Life***?" (Schrödinger 1944). Well, I believe there is a (relatively) ***short answer*** to the good Professor's query; it resides in the life with which we are familiar; the Birds and the Bees, the Dogs and the Cats, Whales and Snails, Lions and Tigers, even Camels, be they of one hump or two, as well as in ourselves.

All these and more exist because of the evolution of a fundamentally, ***inherently unstable*** Microbiological substance known as **DNA**, that ***must replicate*** itself because, in mathematical terms, (*see above*) it ***satisfies its own eigenvalue equation*** or, as ***engineers*** might say; "it exhibits ***positive feedback***," something generally abhorred in feedback systems! Once begun, the process proceeds, "Willy-Nilly", to the creation of another member of a species unless ***aborted*** by the presence of a malfunction or a deliberate act, something we don't discuss herein.

I don't know how bona fide microbiologists might feel about it but for me it's a bit of a Cosmological joke that we, as well as our fellow creatures, require the existence of "an ***inherently unstable*** microbiological process" in order to be here at all! Lot's to think about, maybe?

So; How about this When does DNA replication stop? Well, the way I understand it, ***it doesn't*** — it's just used in a different way! The reason being that "Living" is a ***Process*** whereby individual cells continually die off and must be replaced — something that requires continual sustenance. Eventually, the replacement process becomes faulty with age and it's just a matter of time before the whole subject becomes of no concern for each of us!

6

A (Particular) Connection to Chemistry

Since DNA is a biochemical entity, in fact a very complex biological molecule, discovering its ***structure*** would most generally be considered an exercise in ***biochemistry*** which, by definition, may be viewed as a branch of ***Chemistry***. However, if you recall, a bit of ***structural*** engineering (featuring ***Complementarity***!) was, in the end, found to be crucial to that discovery! Conversely, it may have occurred to you that all those geometric, structural illustrations we used in this book to describe the development of a ***taxonomy*** for our ***Alternative Model*** of the Elementary Particles would appear to have a distinctly ***chemical*** flavor! On the other hand, ***Chemistry*** itself, generally speaking, is closely involved with ***geometric***, ***structural*** considerations; what we are usually concerned with is how a set of ***atoms*** are deployed and connected together to construct a particular ***molecule***.

Hence, the publication of a book by ***Erica Flapan*** (Flapan 2000) entitled "When Topology Meets Chemistry" with the subtitle "A Topological Look At Molecular Chirality" is of particular interest to what we are about to talk about in this very short chapter. Chapter 1 in Flapan's book is entitled "***Stereochemical Topology***" and the first sentence says "Stereochemistry is the study of the three-dimensional structure of molecules and topology is the study of those properties of geometrical objects that are invariant under continuous transformations."

The notion of ***chirality*** is also described in the book thusly: "a molecule that is distinct from its mirror image is said to be ***chiral***, whereas one that can chemically change itself into its mirror image is said to be ***achiral***". The word comes from the ancient Greek word *cheir* which means "hand". Which makes sense and may also bring to mind the notion in physics of "***parity***" and specifically, of the mirror image of the products of Beta decay, or more specifically, the spin of the neutrinos that emerge (which we shall not attempt to discuss in detail at this point!).

On the other hand, it turns out that our ***Alternative Model*** can, in a sense, be said to be directly germane to ***Flapan's*** discussion of chirality that begins, with the notion of a "***Moebius ladder***". This is a large molecule where the ***rungs***, which may consist of a short sequence of small molecules, that can cross each other thus creating a ***knot***. Flapan shows a picture on page 7 of the book of such a molecule (first synthesized by ***Dietrich-Buchecker*** and ***Sauvage*** in 1989) and three pages later the juxtaposition of that molecule and its mirror image.

What those pictures show (take my word for it) are the molecule's overall nature is that of the ***trefoil*** as evidenced by the three ***crossovers***, the ***trefoil*** being of course one of the two torus knots (replicated by both left and right twist) whose concatenation into ***Moebius Strips (MS)*** constitutes the basis for our Alternative Model's taxonomy. As we saw in Chap. 3, the two trefoil (three half-twist) MS form an ***achiral*** pair as the two single half-twist MS (we recall that the NHT of the MS is the same as the number of crossings of the associated knot). The ***chirality*** (or lack thereof) of a knot or the associated MS is shown most readily by beginning with the associated braid structure. So, that's all we have to say about the ***knot*** (and by extension the particle) ***connection*** to ***Biochemistry*** (and by extension) to ***Chemistry*** although, beginning about several decades ago, the use of knot theory in chemistry has become increasingly widespread.

II
Some Basic Physics

God in the beginning formed matter in solid, massy, hard, impenetrable, movable particles of such sizes and figures, and which such other properties, and such proportions to space, as most conduced to the end for which he formed them.

Isaac Newton (1730)

One must be wary of making unwarranted assumptions (per our Preface), and Isaac's was a whopper.

Jack Avrin (2016)

This Section deals with some aspects of fundamental Physics that are important not only because they were ground-breaking achievements but because they unify aspects of physics whose development had previously proceeded along separate paths. Furthermore, as we shall see, the unification in each case may be viewed as another aspect of the Principle of Complementarity in action.

7

Dynamics

This is the place to be basic; maybe quite basic. But not too basic; for one thing, in keeping with the purpose of this book, I have an ulterior motive; I'm really mainly interested in the ***complementarity*** exhibited by the formalisms we deal with. Also, we ought to agree on some things to start with or else we shall never get to the end of this chapter. To begin with, we should know what to call this chapter. I, personally, got stuck on whether to call it ***Dynamics*** or ***Mechanics***. So, I looked in the dictionary and guess what: it looks like the dictionary couldn't make up its mind either — something vague about "forces" acting on "things".

And then, I read here and there that Mechanics includes both Statics and Dynamics. So, I decided to go with ***Dynamics*** mainly because I don't think we care about Statics herein and also that way, it seemed easier to segue into ***Thermodynamics*** or ***Aerodynamics***, etc. later on.

But, then I noticed that some of the revered, ancient founders of that theory invoked the action of "***Mechanisms***". And then there was ***Whittaker*** (1989) in the land of the Brits who (in my imagination) spake thusly: "In the beginning there was naught upon the land and the sea and in the air but "***kinematics***". There was "**motion**" but not yet was there "***force***" to cause the motion, nor to alter its course. And then there was the land of the Soviets and the also ancient, but not quite so much, inhabitants thereof, ***Landau*** and ***Lifschits*** (1969), who (similarly) proclaimed, "in the beginning there were "***particles***", teeny-weeny particles that took up no room to speak of and that one could push around in various ways (ways more basic than the standard manner in which they themselves were pushed!)". End of imaginary meandering.

All very confusing. So, finally, I decided to strike out on my own like so: In the beginning there were, indeed concepts like "***space***" and "***time***" and "***bodies***" of one kind or another with various attributes that we shall invoke as needed. And there was ***Galileo Galilei***, who invoked the notion of "***velocity***" as the unchanging motion through space and time such that small increments of fixed ***size*** in the former would continue unaltered during small increments of fixed ***duration*** in the latter (Machamer and Miller 2021).

And, noted Signor ***Galilei***, a ponderous body's ***velocity*** would remain unchanging in a straight line unless the body was acted upon by an ***external influence*** called a "***force***", adding that the continuous, constant increase or decrease of a body's ***velocity*** during each sequence of increments with fixed duration. This was to be designated as the body's "***acceleration***", a radical notion, in contradiction to the dictum of ***Aristotle***, the ancient fount of all knowledge (according to the ecclesiastical authorities of the day), what Ari said was that ***force*** produced **velocity**! (Which is not as dumb as it might sound but I don't want to get into things like "impulse".)

Finally, Galileo noted that the ***force*** required to impart a given ***acceleration*** to a given body that experienced some ***reluctance to comply*** which he ascribed to the body's "***inertia***". But then, as it is well-known, his apocryphal[1] experimentation in dropping bodies of different "***weights***" from a height showed that they all fell with the same ***acceleration***. Which told him, that their ***inertia*** was somehow proportional to that ***weight*** as compared to, say, a standard in a balance. In modern terms, that ***inertial mass*** is proportional to ***gravitational mass***.

All this being well before there was an ***Einstein*** (Famous Scientists 2014) or a ***Minkowski*** (O'Connor and Robertson 2015) or even a ***Lagrange*** (Famous Scientists 2021) or a ***Hamilton*** (Wilkins

[1]Actually, what he did was very much like what they do in freshman physics labs with inclined planes, etc.

2005) or others of their ilk. But there did appear a ***Newton*** who proclaimed, from his "perch atop the shoulders of giants" (BBC 2021)[2], that the ***force*** required to impart a given ***acceleration*** to a body of a given "***mass***" was proportional to that mass; in fact, that, Ta Da!

$$\boldsymbol{F} = \boldsymbol{ma} \tag{7-1}$$

is an equation that portended to change the world. But what was that "mass"? It was a couple of centuries before ***Albert Einstein*** announced, even before imprinted T-shirts were invented, that a body's *energy* was proportional to its mass, the proportionality constant being the square of the velocity of light — that is, that $E = mc^2$ and, further, that the "***force***" of ***gravitation*** was really due to the ***curvature*** of space induced by ***mass***, (for example, the reason the planets revolve around the sun is that they experience the curvature of space due to solar mass) thus closing the (mechanical) logical loop.

Well, $f = ma$ and its various manifestations did change the world, not only in our ability to analyze dynamic situations but to accelerate the pace of technological advancement. Of course, the original statement soon began increasing ***sophistication***. But before we get into that, let's take another look at Eq. (7-1). It looks so simple, it's hard to imagine the host of ramifications attributed to its employment. But ramifications aside, note that there are three ways to re-express the equation by viewing each of the terms in turn as a ***parameter*** in which case the remaining terms constitute a ***complementary*** pair (we can also change the order of occurrence of the terms but that makes no difference). Figure 7.1 shows these cases:

1. $\boldsymbol{F}$ is the parameter and, correspondingly, we see $\boldsymbol{m}$ and $\boldsymbol{a}$ varying as a set of hyperbolic ***complementary*** pairs.
2. $\boldsymbol{m}$ is the parameter and $\boldsymbol{F}$ and $\boldsymbol{a}$ vary as a fan of linearly increasing ***complementary*** pairs.
3. $\boldsymbol{a}$ is the parameter and $\boldsymbol{m}$ and $\boldsymbol{F}$ vary as a fan of linearly increasing ***complementary*** pairs.

[2]Do you recall Newton's explanation of how he was able to see farther than most men?

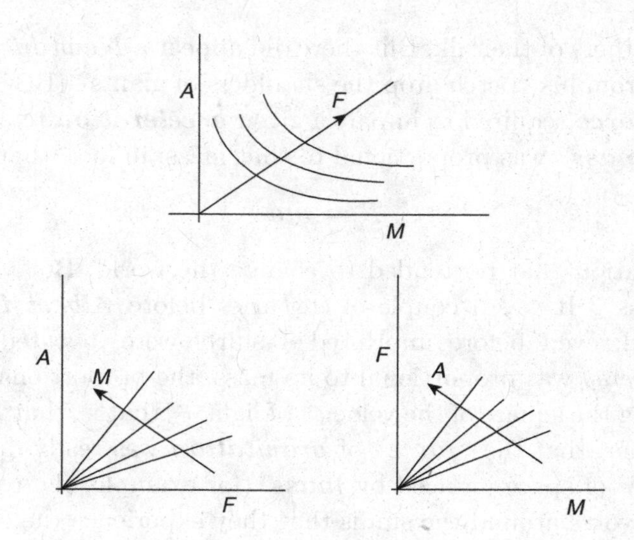

Fig. 7.1. Three Ways to Employ $f = ma$

In practice, the indicated parameter might select ***one*** of the indicated set of characteristics such that, for example, case #2 may be recognized as encompassing that of dropping a weight from a height in order to see how its acceleration varies with gravitational parameters. We can write the equation for ***gravitational force*** as (Another masterful Newtonian proclamation!)

$$F_G = Mm_G/GR^2 \tag{7-2}$$

where G is the gravitational constant, R is the distance from the ***source***, M, of the gravitational force (in the case of experiments on or near the surface of the earth this force is essentially constant) and m_G means "Gravitational mass". We should also write another equation for "inertial" force (subscript I means "inertial") in keeping with the master's famous dictum ($f = ma$)

$$F_I = m_I a_I. \tag{7-3}$$

However, the body being dropped doesn't feel ***any force*** at all, a well-known common experience (see Chap. 11 about how that common experience relates to General Relativity) so we can equate the two forces whereupon, solving for acceleration, we have

$$a_1 = a_G = (M/G)(m_G/m_I)/R^2. \tag{7-4}$$

Furthermore, since all measurements and experimentation indicate that "***inertial***" and "***gravitational***" ***masses*** are ***identical,*** ($m_G/m_L = 1$), we see that the ***acceleration*** due to the ***gravitational effect*** of a large body on a small body ***depends only*** on the large bodies mass and the ***square*** of the distance of the small body from the center of the large one, which near the surface of the Earth is essentially constant.

But, back to that "***sophistication***" we alluded to above: a number of formulations of dynamics evolved in the 18th and 19th century, becoming indispensable ***tools*** for application to a broad range of phenomena especially to elementary particle physics and eventually as adapted to quantum mechanics. One that lots of people use can be summarized in one sentence and all that's left to discuss are the details (which, as is usual, with tools, require some familiarity in their use!). Here it is:

The "sophistication" in this case amounts, first, to the definition of the ***Lagrangian Function*** as $\boldsymbol{L} = \boldsymbol{T} - \boldsymbol{V}$, ($T$ and V being kinetic and potential energy, respectively), its integral, known as the "***Action Integral***", ***Hamilton's principle*** regarding the ***minimization*** thereof in "***Generalized Coordinates***" and, finally, the ***Euler-Lagrange equations*** that result and can be applied to a ***generalized system of coordinates*** in order to determine the dynamic behavior of a wide variety of phenomena[3].

At which point, depending on your background, you might be inclined to protest "Wait a minute"; what's that all about and who are those guys? And you would be justified because we have left out something in this story, namely another ***Newtonian*** invention: the ***Calculus***!

That's right, ***Differential Calculus*** to compute the rates of change of known functions for example, the instantaneous ***distance***

---

[3]Note the ***complementarity*** of T and V in the Lagrangian function and consequently, the Action Integral.

from the point of release of an object dropped from the tower of Pisa versus ***time*** — in other words, its ***velocity***. Or conversely, ***Integral Calculus*** to calculate, for example, the total velocity attained by a rocket based on accelerometer readings. The present advanced state of the science of Dynamics would have been impossible without the development of the ***calculus***, meaning ***both*** Differential and Integral Calculus. The physics and the mathematics are synonymous and developed together. Moreover, those two varieties of the calculus are really just two ***complementary*** aspects of the fundamental way to analyze physical processes; both are necessary tools.

By the way, before we go back to talk about those icons of "sophistication" mentioned above, it's of interest to note that the renowned Philosopher-Mathematician, Leibniz[4], in Germany, also came up with the notion of the calculus at about the same time as Newton (Look 2013) but expressed with somewhat more felicitous notation, namely the ***integral*** sign we use today and the dx/dt notation for the ***derivative*** of $x(t)$ versus t, where Newton used the dot over the x and so forth. Actually, as is well known, both notations are in general use, often, as we shall see below, occurring in the same expression!

So, now, about ***Hamilton's principle***, here's what ***Cornelius Lanczos*** says about it in his excellent book entitled "The Variational Principles of Mechanics (Lanczsos 1949): "Hamilton's principle which states that the motion of an arbitrary mechanical system occurs in such a way that the definite integral A becomes stationary for arbitrary possible variations of the configuration of the system, provided the initial and final configurations of the system are prescribed", where the definite integral alluded to is the ***action integral*** of the Lagrangian mentioned above.

At this point, it occurred to me that perhaps the best way to illustrate the basic dynamics in both Newtonian and Hamiltonian terms was ***with an example*** so here it is: the dynamics of a simple pendulum. Actually, I borrowed this from my previous book where, although it's well known, I wrote it down therein as an example of a

[4]Gottfried Wilhelm Leibniz, (1646–1716).

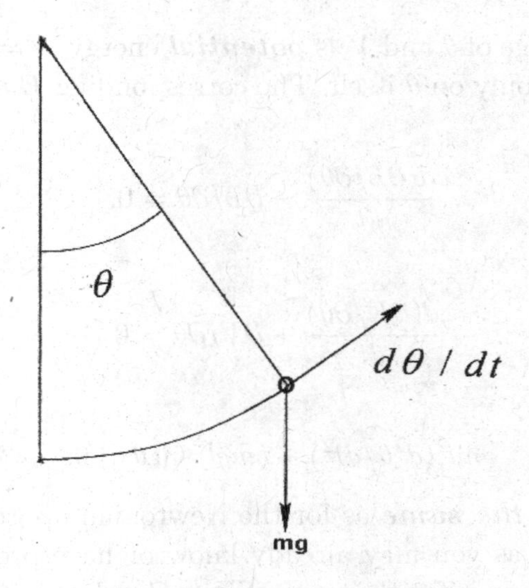

Fig. 7.2. Pendulum

so-called "***Sine-Gordon***" system that I could use for the solitonic behavior of my model of the elementary particles. Figure 7.2 depicts an idealized pendulum consisting of a stiff, weightless rod of length l and a weight of mass m constrained to rotate in a vertical plane under the influence of gravity.

The behavior of this elementary device is readily derived via straightforward Newtonian ($F = ma$) mechanics but specialized, here, to ***angular*** coordinates where the restoring ***torque***, $[-(mgl)\sin\theta]$, equals the ***moment of inertia***, ml^2, multiplied by the angular ***acceleration***, $d^2\theta/dt^2$, i.e.

$$ml^2(d^2\theta/dt^2) = -(mgl)\sin\theta, \tag{7-5}$$

or, in ***Sine-Gordon*** format,

$$(d^2\theta/dt^2) + (g/l)\sin\theta = 0, \tag{7-6}$$

an expression that can also be derived from a ***Lagrangian***, $L = K - V$, in a relatively straightforward way, where, generally speaking, potential energy V is a function only of an extensive variable and kinetic energy is a function only of its derivative. Thus here, ***kinetic*** energy is given by $K = (1/2)ml^2\dot{\theta}^2$, which, obviously, depends only on the rate of change of θ and V is ***potential*** energy, $V = mg(l - l\cos\theta)$ which depends only on θ itself. The corresponding ***Euler-Lagrange*** equation,

$$\frac{d(\partial L/\partial\theta)}{dt} - \partial L/\partial\theta = 0, \tag{7-7}$$

then becomes

$$\frac{d(\partial K/\partial\dot{\theta})}{dt} + \partial V/\partial\theta = 0. \tag{7-8}$$

That is

$$ml^2(d^2\theta/dt^2) + (mgl)\sin\theta = 0, \tag{7-9}$$

in other words, ***the same*** as for the Newtonian approach.

By the way, as you may already know or have probably already surmised, the reason for the name ***Sine-Gordon*** is the presence of the sine function of the angle rather than just the angle itself as in the usual discussion of simple harmonic motion for the case of small-amplitude pendulum swings. Also, the second term in Eq. (7-7), $\partial L/\partial\theta$ is a force that changes the momentum $\partial L/\partial\dot{\theta}$. In the general case, angle θ is replaced with an appropriate generalized variable.

Of course, Hamilton's ***principle*** is not the only way to approach fundamental dynamics: a couple of well-known others are: 1. Hamilton's ***equations*** that refer to the ***Hamiltonian*** of the system (That's right; it's the same guy!), a kind of generalized ***energy***, are known as 2. ***Poisson Brackets*** that when applied to the Hamiltonian lead to a statement of the ***conservation of energy***. I shall just state the essential formalism here, going no further except, looking ahead, to note that the Brackets are an essential starting point for one particular concept upon which to base ***Quantum Mechanics*** (See Sec. 1V).

So, the ***Hamiltonian*** of the complementary pair ***position*** q and ***momentum*** p is

$$H(q,p) = \sum_{\beta=1}^{n} p_\beta \dot{q}_\beta(q,p) - L\{q, \dot{q}(q,p)\} \tag{7-10}$$

whereupon, ***Hamilton's equations*** are the ***complementary*** pair

$$\partial H/\partial p_\alpha = \dot{q}_i \text{ and}$$
$$\partial H/\partial q_\alpha = -\dot{p}_i. \tag{7-11}$$

Meanwhile, the ***Poisson Bracket*** of two variables f and g is defined as

$$[f,g] = \sum_k \left(\frac{\partial f}{\partial p_k}\frac{\partial g}{\partial q_k} - \frac{\partial f}{\partial q_k}\frac{\partial g}{\partial p_k} \right).$$

(Note the ***complementarity*** of their appearance here).

So now, suppose f is the ***Hamiltonian***: then, g is known as an "***integral of the motion***" of the system under consideration; that is, its ***total*** time derivative vanishes, i.e.

$$\frac{dg}{dt} = \frac{\partial g}{\partial t} + [H,g] = 0 \tag{7-12}$$

whereupon, if it is also ***not*** an ***explicit*** function of time, it must satisfy $[H,g] = 0$.

Which is about all we have to say regarding elementary dynamics at this time, except for one commentary concerning a somewhat controversial subject therein that was brought to my attention recently by an excerpt in a book by Dwight E Neuenschwander (Neuenschwander 2011) entitled "Emmy Noether's Wonderful Theorem," a theorem that, the author asserts, "forms an organizing principle for all of physics" and that we shall meet in the next chapter. The book is very thorough with a large part devoted to developing a theoretical basis upon which to emplace the theorem itself including fundamental notions like the Euler-Lagrange equation, Hamilton's and Fermat's principles and the Action integral in which, you may recall, the integrand is the ***difference*** between ***kinetic*** and ***potential energies***.

At that point in his book, Neuenschwander raises the rhetorical question as to why that's the case — that is, why it should be the ***difference*** and not the ***sum*** (you may have puzzled about it yourself; I certainly did) and, as an aside, he takes the time to indicate various avenues for answering the question. Well, upon reading that part of the book I spent some time ruminating about the question myself and have come to the following conclusion: in at least one situation, the question doesn't really apply — ***both*** versions are applicable! And, in fact, they are ***complementary***! I was going to say the question is nugatory because I recently learned that word but, fortunately, I looked it up again and the definition was "trifling, vain, futile, insignificant, worthless", none of which are fair; it's really a perfectly valid, important question discussed in some detail in Chap. 11.

8

Thermodynamics

The science of ***Thermodynamics*** is a voluminous subject, wide and deep and at times not easily traversed by those used to straightforward, well-defined technical description. I am reminded of the Introduction to this book wherein I talked about *The Meaning of "is"* as a barrier, a thicket that bars our way to ultimate understanding. Nevertheless, we shall proceed, wielding our Machete of Meaning to eliminate the polysemic (!) thickets that impede our progress to thermodynamic understanding (using flowery words and phrases to bolster our courage!).

Historically, Thermodynamics started slowly but really picked up steam (so to speak!) in parallel with the Industrial Revolution especially during the latter part of the 18th century and throughout the 19th as the use of larger and larger machinery, consuming greater and greater amounts of fuel became more prevalent. In contrast to the science of Mechanics for which, beginning with Galileo and Newton, theory and experiment developed more or less hand-in-hand, Thermodynamic science developed mainly in an empirical rather than a theoretical way with a lot of experimentation on ***macroscopic*** bodies or media but little more than speculation as to their likely ***microscopic*** nature. (See, for example, Bozsaky 2010.)

Of course, the body of knowledge accumulated had to be organized and that led to the introduction of a set of not always easily-defined basic concepts and what are known today as the ***Laws of Thermodynamics***. Eventually, also, that speculation about what was inside those "macroscopes" developed into probabilistic modeling and to what became known as the ***Statistical Mechanics***

(SM) of bodies or media composed of large ***ensembles*** of "particles" of one type or another for which the validity of a statistical point of view could be justified by the way in which it explained the macroscopic thermodynamics.

With luck, we'll get into SM theory eventually but right now I want to talk about Thermodynamics at a relatively elementary level (my comfort zone!) because the ***complementarity*** of the elements of the theory in each case is really quite apparent therein. Furthermore, the aforesaid laws have been found to be applicable over a very large range of phenomena and "***phases***" of matter: solid, liquid, gaseous and various variations and combinations of each as long as the "ensemble" model is valid.

The canonical example is of course just water (LibreTexts 2021) which can exist, even coexist in all three phases, as shown in Fig. 8.1 drawn to show three pressure vs. temperature curves that separate them, two at a time, and meet in what is commonly known as (surprise!) the "***triple point***". As we see, the "melting" curve rises steeply to separate water from ice (that's where Global Warming is turning the Artic into slush) and the curve that continues to rise to the right with temperature separating water from steam and is what we experience as we heat the water for our morning coffee. (Actually, there is a parameter known as the "heat of vaporization" whose

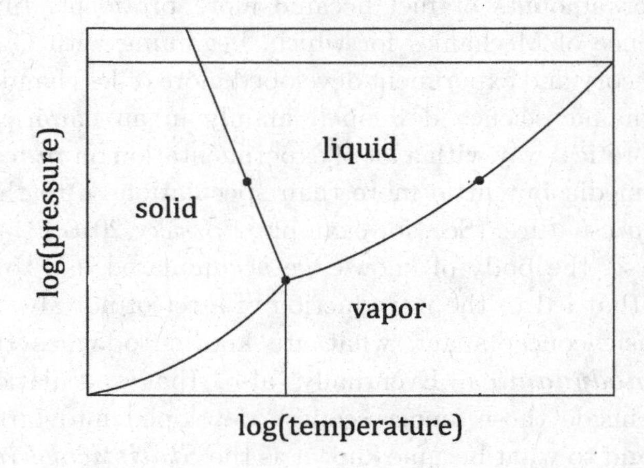

Fig. 8.1. Phases of Water.

energy equivalent is the amount of energy a water molecule needs to free itself from the unbalanced force that otherwise sequesters it to the surface.)

Discrete models have been developed for each such "phase". In what follows we shall focus on ***steam*** for which the canonical model is the ***equation of state*** for gaseous matter, colloquially, "the gas law"; you may remember it from first year college or perhaps even high school physics:

$$PV = nRT. \tag{8-1}$$

One reason to bring it up is that it's an example of the kind of two-at-a-time ***complementarity*** exhibited by the ***triplet***, pressure, volume and temperature for many Thermodynamic systems. (just like that $f = ma$ triplet in dynamics). Another reason[1] is that we can use that equation to make a rather counterculture point: one can maintain that the name shouldn't really be ***Thermodynamics***, the science of "***Therms***", the measure of ***heat***; it should be ***Ergodynamics***, the science of "***Ergs***", the measure of "***energy***". It's easy to demonstrate:

Consider a volume, V, of gas at some temperature, T, and under some pressure, P. On the ***left*** of the equation we have ***Pressure*** which is ***force*** per unit ***area*** multiplied by ***Volume*** which produces ***force*** times ***length***, which is ***Work***, a form of ***Energy***! And on the ***right***, we have, n, which is the number of ***moles***, a mole being a number equal to the molecular weight of the substance of interest expressed in grams (I think) multiplied by ***Temperature*** and by the so-called ***gas constant***, R, that converts ***temperature*** into ***energy*** in ***ergs per mole***. Ergo: (automatic pun!) both sides agree: it's all about ***energy***! By the way, this is the first time the combination RT, another example of complementarity, has come up herein; look for it in what follows!

In any event, I must again call upon the authority of the legendary Professor ***Enrico Fermi*** to encapsulate the subject which he does

[1]Actually, there is a third, very important reason related to reason #1.

in the introduction to his book (entitled, fittingly and to the point), as "***Thermodynamics***" (Fermi 1937), which begins with the definition, "Thermodynamics is concerned with the transformation of heat into mechanical work and the opposite transformation of mechanical work into heat." It goes on to note that it is only in "comparatively recent times" that heat was recognized as a "form of energy that can be changed into other forms", citing the discovery, in 1842, of the equivalence of heat and mechanical work by R. J. Mayer who "...made the first announcement of the principle of the ***conservation of energy***; the ***first law*** of thermodynamics" (Fermi's book was published in 1936).

So, there we have it, the ***first law***, or at least a highly simplified version of it that we might relate to the following scenario: Suppose we have a mechanism, say a steam engine, to which you supply a quantity of heat, Q, say by turning some water in a boiler to steam which expands, thus performing an amount of work, $\boldsymbol{W}$, say by forcing the steam into a cylinder to drive a piston that ends up, via some mechanical linkage, driving a propeller mechanism propelling a ship upon the water or a set of driving wheels propelling a train locomotive along the track. (The great thing about the locomotive is that you can actually see the linkage cranking the drive wheels!)

But then, inexorably, as the wheel keeps turning everything returns to the initial condition except that we have expended some energy

$$\boldsymbol{U} = \boldsymbol{Q} - \boldsymbol{W}, \tag{8-2}$$

where $\boldsymbol{W}$ is the ***work*** the system actually performs in doing all that driving we described above. And, we might note, as per this equation as well as Fermi's definition, work and heat are ***complementary*** concepts.

So, in more detail, here's what the ***first law*** says: in the first place, $\boldsymbol{U}$ has to be positive; you're supposed to be ***supplying*** energy ***to*** the system, not drawing energy ***from*** it implying that $\boldsymbol{W}$ can't be larger than Q. In fact, it can't even be exactly equal to it, thus proscribing the feasibility of a ***perpetual motion*** machine that uses up no energy at all. If you want to keep on driving that ship through the water, or the train on the tracks at a constant velocity, you have to keep on supplying heat energy on every cycle, in effect continuously. That's what the law says and abiding by it was something that provided employment for generations of brawny coal "stokers". (Nowadays of course they just open the valve on an oil tank.) By the way, all that activity we described above must occur upon every stroke of the piston in order for the ship or the train to get where it's going — ***complementary teamwork*** in action.

In practice, that extra energy ends up increasing pressure in the boiler so that every once in a while, the engineer has to "let off steam" and you can hear "that lonesome whistle[2] coming down the trestle". As a boy, lying awake in bed, I heard it even though the railroad tracks were a long way away. In any event, the basic message of the first law is that if you're trying to do some work, you always put in ***more energy*** in the form of heat than you get out in the form of work. That's why you get hot and sweaty when you exercise or do physical work; you're letting off steam! (Actually, you're getting rid of heat in the form of "the heat of vaporization" we mentioned above that converts droplets of liquid into vapor.)

But, it gets worse; the first law didn't tell the whole story! For the rest, we need the ***Second law*** and for that we need to thank the French Army or, more explicitly, Nicolas Leonard ***Sadi-Carnot***, a Military Engineer therein who, in 1824 published a book entitled (English translation) "Reflections on the Motive Power of Fire" (Carnot, Clapeyron and Clausius 1960) that many view as constituting the ***foundation*** of the second law and, indeed of the breakthrough that led to the concept of ***Entropy*** and the beginning of the modern era in Thermodynamics. The book documented some penetrating analysis Carnot undertook in order to make sense out of what, in his day, was involved in the confusing trial-and-error, ad hoc world of mechanisms that relied on heat to do work.

But before we become immersed in the second law, we must not forget that, actually, there exists what became known as the "Zeroth Law", something that was actually added after the others

[2]Remember that song? It was called "The blues in the night".

had been laid down upon the land. It says that if two "systems" are each in "***thermal equilibrium***", with a ***third*** "system" they are also in thermal equilibrium with ***each other***. It occurred to me that by the time mentioned above (1842), it was long known that ***heat*** was something that could be subjected to measurement in terms of its ***temperature*** relative to a standard, for example, by a "***thermometer***" that, for example, absorbs heat that does ***work*** to linearly expand some Mercury in a graduated cylindrical tube.

When, for some reason you want to know your temperature, you stick such a thermometer in your mouth and wait until the reading settles down (indicating ***thermal equilibrium***) to read what the thermometer says. Under the assumption that your temperature hasn't changed you can then check out (***calibrate***) another thermometer you were not certain about in the same way by comparison with the first one's reading (since, according to the zeroth law, the two thermometers are thereby in thermal equilibrium with each other). And likewise, the air ***pressure*** in your automobile's tires can be measured by letting some of it ***push*** a tiny cylinder along a graduated surface in a "***pressure gauge***", and you can likewise calibrate another gauge as in the above. In fact, from a practical point of view, the zeroth law is of value mainly for just that kind of activity.

And, going back to the First Law for a moment; given the triumvirate P, V and T, it is possible, at least for systems such as gases and simple fluids, to draw three kinds of "***phase diagrams***" each of which shows how any ***pair*** of the triumvirate varies in ***two dimensions***, with the third viewed as a ***parameter***. For example, in the above gas law, fixing T shows that P and V vary as a hyperbolic "***Isotherm***", and fixing either P or V implies families of expanding fanlike characteristics: the situation is just like the one we talked about in the $f = ma$ dynamic chapter. But, in any case, we always have a **pair** in a parametric, give-and-take ***complementary*** relationship.

Anyway, back to Carnot and here's what he had in mind: in the first place he was motivated by the idea that what was lacking in our ability to systemize thermodynamics was some kind of ***ideal***

behavior for a closed system — i.e. one that returned to its initial condition without losing the energy it started with, in other words, a ***reference*** that indicates the best that can be done. What he came up with came to be called, not surprisingly, the ***Carnot cycle***, an idealistic system impossible to duplicate in practice but one whose performance could, in some sense, be quantified and serve as standard against which other systems could be compared.

As it turned out, the ***second law*** of thermodynamics emerged as a result of analyzing the behavior of that cycle. To explain the law in those terms, we need to go back to our friendly ***gas law*** and avail ourselves to one of those phase diagrams, namely the P versus V diagram, that depicts a ***Carnot Cycle*** as shown in Fig. 8.2 below.

Our explanation involves nothing new or novel; after all these ideas have been around since 1824! As we see, there are two curves, more horizontal than vertical, known as ***isotherms*** and two, steeper, so-called **adiabatic** curves that implement the transition from one isotherm to the other. The arrows indicate a closed circuit starting at point A and proceeding ***isothermally*** to point B (which is greater in volume but somewhat lower in pressure) both on the upper curve, then ***adiabatically*** down to the lower isotherm at C, back **isothermally** to D and finally back up ***adiabatically*** to the starting point at A.

Of course, I don't have to mention that ***"Isothermally"*** means "without changing temperature" and "***adiabatically***" means

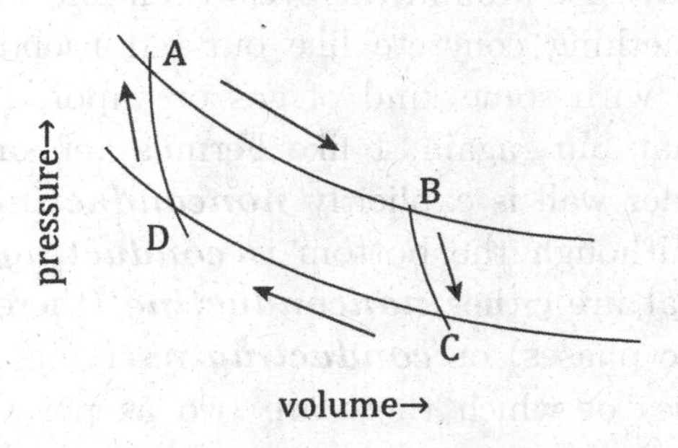

Fig. 8.2. Carnot Cycle.

without absorbing or emitting energy of any kind from the surroundings. As per the gas law, the isotherms are sections of hyperbolas ($pV =$ a constant) and, although it doesn't look like it, the "adiabats" are "qualitatively" (but not exactly) similar. As they say, "It can be shown", in fact, Fermi shows that the operative characteristic therein is

$$pV^K = \text{a constant} \tag{8-3}$$

where the exponent, K, is the ratio of a couple of what are known as ***specific heats***, that is the rate of temperature rise with respect to heat addition in the case of constant volume on the one hand, and constant pressure on the other.

I shall forgo elucidation on this point so as not to lose focus, but the net result is that the exponent, is what makes the adiabatic transitions between the isotherms rather steep in comparison with the isotherm characteristic itself (as we see in Fig. 8.2). However, I can explain rather quickly why that is the case: you just form the derivative dp/dv in each of the two formulas, namely

$$\begin{aligned} pV &= \text{a constant, and} \\ pV^K &= \text{a constant} \end{aligned} \tag{8-4}$$

whereupon we find the slope in the second case to be greater by a factor of K.

Nevertheless, to properly elucidate Carnot cycle behavior we really need to navigate around the cycle itself and for that, it helps to think of something concrete like our old friend the piston in a cylinder filled with some kind of gas or vapor. Apparently lots of people do that but, again, I like Fermi's version which is one where the cylinder wall is explicitly ***nonconducting*** and so is its' top. However, although the bottom is ***conducting*** it can be ***set*** on ***surfaces*** that are either ***nonconducting*** (there's one of those for the adiabatic phases) or ***conducting*** reservoirs of heat for the isothermal phases of which there are two as per the description, above.

So, with reference again to Fig. 8.2 that's what happens in the four cases involved in sequence here: the progression[3] from A to B proceeds ***isothermally***; then from B to C ***adiabatically***, from C to D again ***isothermally***, and finally back up from D to A again, ***adiabatically***. In the first two phases the pressure in the cylinder must automatically ***decrease*** as the contents ***expand***. To implement that, we ***pull*** on the piston, very, very slowly and carefully (to ensure that the process is ***reversible***), not exerting any force ("to speak of"). In the second two, the pressure must ***increase*** as the contents compress and to implement that we ***similarly push*** on the piston.

And here's the crux of the whole cycle — ***heat transfer***! The transfer of heat from a higher temperature reservoir to one at a lower temperature. As per our model, in traversing the upper isotherm, a certain quantity of heat, ***Q1***, is transferred ***to*** the cylinder ***from*** the hotter reservoir. Then, in traversing the lower isotherm a ***second*** amount of heat is transferred ***from*** the cylinder ***to*** the cooler reservoir; let's call it $-Q2$, negative because it is removed ***from*** the cylinder. The total amount of heat ***absorbed by our cylinder*** is therefore ***Q1 – Q2***, which, according to the ***First Law*** for a complete cycle in which *there is* ***no loss of total energy***, is equal to the amount of ***work*** done by the system. That is to say

$$W = Q1 - Q2. \tag{8-5}$$

What we're looking at here is the Robin Hood of Thermodynamics! If *Thermodynamics* were money, the Carnot cycle would be taking from the *rich* ***reservoir*** to give to the poor one. If it were part of an economic system, it would be called a "Share the wealth plan". We are fortunate to have at our disposal those two heat ***reservoirs*** that are essentially unaffected by the cycle but then the Carnot cycle is not required to adhere to the limitations of reality. In fact, it would appear that, at this point, there is no immediately apparent

[3]Actually, no one ever explains what makes the system do the "progressing". Presumably "something" gets it started and then the rest is automatically cyclic.

reason why, in principle, the cycle could not continue on down the temperature scale (but see below).

And speaking of cycles, the Carnot cycle works because it ***is*** a cycle. And the reason it is a cycle is because of its' ***complementarity***. The pair of ***isotherms*** and the pair of ***adiabatics complement*** each other. Furthermore, the two ***isotherms complement*** each other, one progressing mainly towards increasing volume and the other towards decreasing volume thus cancelling each other out (as far as volume is concerned). And similarly, the two adiabatics complement each other mainly in the pressure dimension.

If all the heat absorbed at the higher temperature were transferred to the lower reservoir the Carnot Cycle would be 100% efficient! The ***efficiency*** of the Carnot Cycle is therefore defined as the ratio of the work performed during the cycle to the heat absorbed at the higher Isotherm, i.e.

$$\eta = \frac{Q1 - Q2}{Q1} = 1 - \frac{Q2}{Q1}. \tag{8-6}$$

By ***directly*** calculating the ***work performed*** we can then translate this expression into an equivalent one in terms of the associated reservoir temperatures:

$$\eta = \frac{T_H - T_L}{T_H} = 1 - \frac{T_L}{T_H} \tag{8-7}$$

where the subscript H and L signify the upper and lower isotherms, respectively.

To show this we go back to the gas law once again which shows ***work performed*** to be expressible as

$$\int pdV = \int (nRT/V)dV. \tag{8-8}$$

Then, since ***work*** is performed ***only*** during the ***isothermal*** stages, this leads to an expression for efficiency as

$$\eta = \frac{-nRT_H \ell n(V_2/V_1) + nRT_L \ell n(V_4/V_3)}{-nRT_H \ell n(V_2/V_1)} \tag{8-9}$$

where the subscript H and L signify the upper and lower isotherms, respectively, and we have redefined points A, B, C, and D, nominally the subscripts in the parenthesis as 1, 2, 3 and 4, respectively.

We also note that, during the ***adiabatic*** stages, temperature and volume trade off so that we can write

$$\frac{T_2}{T_3} = \frac{V_3}{V_2} \text{ and } \frac{T_1}{T_4} = \frac{V_4}{V_1}, \tag{8-10}$$

which, (one more realization!), since $T_1 = T_2$ and $T_3 = T_4$, gives us

$$\frac{V_4}{V_3} = \frac{V_1}{V_2}, \tag{8-11}$$

$$\eta = \frac{nRT_H\ell n(V_2/V_1) - nRT_L\ell n(V_2/V_1)}{nRT_H\ell n(V_2/V_1)}, \tag{8-12}$$

in which, since everything cancels except the temperatures, produces the indicated expression, Eq. (8-7), above that we repeat here

$$\eta = \frac{T_H - T_L}{T_H} = 1 - \frac{T_L}{T_H}.$$

Evidently, efficiency is promoted by operating at as ***low*** a ***lower*** temperature, T_L, as possible. However, since how low that can get is often quite limited, operating at as ***high*** a ***higher*** temperature, T_H, as possible which substituted into Eq. (8-9), gives us (using $\ell n(b/a) = -\ell n(a/b)$) is desirable.

In any event, as we shall see, all this talk about the importance of temperature brings up the subject of ***Entropy***! The concept of entropy was introduced by Rudolph ***Clausius*** in 1850 (Bozsaky 2010), then a Professor of physics at the Royal Artillery and Engineering School in Berlin, and an admirer of Carnot and his accomplishment (Carnot, Clapeyron and Clausius 1960). Nevertheless in an 1862 paper, he took a bit of issue with the way Carnot had summarized the main message of his cycle, basically the original statement of the second law.

Here's what Carnot had said (***Carnot's Theorem***): "No engine operating between two heat reservoirs can be more efficient than a Carnot engine operating between those same reservoirs". And in a corollary: "All reversible engines operating between the same reservoirs are equally efficient" (Carnot, Clapeyron and Clausius 1960).

And what ***Clausius*** said was "The algebraic sum of all transformations occurring in a cyclic process can only be positive, or, as an extreme case, equal to nothing (Carnot, Clapeyron and Clausius 1960)". What he had in mind was a ***mathematical expression*** $\int \frac{dQ}{T}$ which he called S and said must be ≥ 0 for the general cyclic process a body undergoes, and with the equality holding only for reversible such processes where, dQ is the heat the body gives up, irreversibly, to any reservoir and T is the absolute temperature of the body when so doing.

It seems to me that the difference between the two statements is basically due to a difference in philosophical point of view with Carnot's being more Epistemological and Clausius' more Ontological. Or in military terms (both men being involved with the military) the difference being that between strategy and tactics, with Carnot the strategist and Clausius the tactician. On the other hand, in 1865 Clausius named the above irreversible quantity as follows "I propose to name the quantity S the entropy of the system after the Greek word [$\tau\rho o\pi\eta$, trope], the transformation. I have deliberately chosen the word entropy to be as similar as possible to the word energy: the two quantities to be named by these are so closely in physical significance that a certain similarity in their names appears to be appropriate" (Carnot, Clapeyron and Clausius 1960).

This indicates to me that he had thought a lot about what was really fundamental in thermodynamics, namely energy, and that his concept of ***Entropy*** represented an ***irreversible loss*** of energy. That is to say, energy **that could no longer be put to work**.

Entropy is a large subject, one that impacts all phases of physics. Some say that on a cosmic level it may be viewed as the most important concept of physics with numerous ways to be expressed especially with regard to its statistical aspects. Some pages back we interrupted Professor Fermi just as he mentioned ***Statistical Mechanics***; that's mechanics but mainly talking about the ***probabilities*** of things happening (rather than actually computing the nature of their occurrence) and we have formalisms such as the ***Maxwell-Boltzmann*** velocity distribution for molecules, the statistical nature of ***Entropy*** and eventually of the subject of "***Information***".

For the most part, Statistical Mechanics will be covered in a chapter so named that has been delayed until almost the end of the book so as to mesh with considerations of a more comprehensive nature. Nevertheless, we can, at this point, introduce something of statistical significance that is readily seen to relate to what we have said so far about ***entropy***, namely the ***Boltzmann*** version of that we shall designate as

$$S_B = k \ln p_B \tag{8-13}$$

where $k = R/A$ is the Boltzmann constant. R is the gas constant we met before and A is Arago's number, the number of molecules in a mole's worth of such so that k converts temperature to energy. According to Fermi once again, p_B is to be regarded as the probability that the ***most stable state*** of the system under consideration has been occupied.

Of course, the question immediately arises as to whether ***Boltzmann's*** version ***is indeed*** a properly ***probabilistic*** way to express ***Clausius***' thermodynamic version, one that correlates with that version and to answer that question we begin with the recognition that entropy is an ***additive*** entity. Thus we consider a thermodynamic system that may be regarded as the combination of two components so that we can write its total entropy as (with C for "Carnot")

$$S_C = S_{C1} + S_{C2}. \tag{8-14}$$

In the meantime we note that ***probability*** is a ***multiplicative*** entity so that, given the occurrence of such in the ***Boltzmann*** version we can also consider a ***hypothetical*** entity, as a tentative surrogate — a "placeholder" — for the Boltzmann version, i.e. (with H for "Hypothetical")

$$S_H = S_{H1} S_{H2} \tag{8-15}$$

and pose the question as to whether

$$S_C = S_H. \tag{8-16}$$

Upon reflection we realize Eq. (8-16) holds ***if and only if*** all quantities on the RHS of Eq. (8-15) are ***logarithmic*** in nature. That is if, say that

$$S_{Ck} = \ell n(Ck) \text{ and}$$
$$S_{Hk} = \ell n(Hk) \tag{8-17}$$

Where $k = 1$ and 2, so that, finally,

$$\ell n(Ck + Hk) = \ell n(Ck) + \ell n(Hk), \tag{8-18}$$

a known ***mathematical fact***, in words, that the ***logarithm of a product*** is equal to the ***sum*** of the ***logarithms*** of the two ***multiplicative factors***. Ergo, the ***Boltzmann*** version featuring a ***logarithm*** is, indeed ***equivalent*** to the classical, ***thermodynamic*** version of ***Clausius***[4].

[4]Some of you may remember Log Tables, perhaps even Slide rules!

9

Energy-wise Comparison of Two Exercises

Believe it or not — you be the judge — this chapter is a logical follow-on to the two preceding chapters because, as you'll see, in essence, it's all about energy. The way the subject came up goes back to a dinner my wife and I were invited to share with a very nice couple across the street. The wide-ranging conversation that ensued had somehow turned to (the subject of) exercise and when I started to relate how I used to be able to perform multiple "Muscle-ups" on the high-bar in days long gone, Tom remarked that it reminded him of a well-known weight lifting procedure. As it turns out, I was also acquainted with that procedure (which, by the way is, descriptively, known as "Clean and Jerk"!) and I quite agreed with him even though the relationship between the two activities might seem far-fetched at first thought (to the uninitiated of course!). At second thought, however, clearly both activities involve hoisting a weight, an actual dead weight in one case and body weight in the other, and both involve grasping a long, horizontal bar with both hands and shifting arm and hand orientation thereto at some phase during each procedure in order to accommodate a change in the method of hoist.

After coming home from our dinner, it occurred to me that there might actually be a new chapter of the book lurking in a more detailed comparison of the two exercises because, while some differences between the two exist, there are indeed some striking similarities and, as it turns out, a not totally surprising connection to ***Complementarity*** that I found most gratifying, something we really have to build up to. Please don't skip to the end of the chapter!

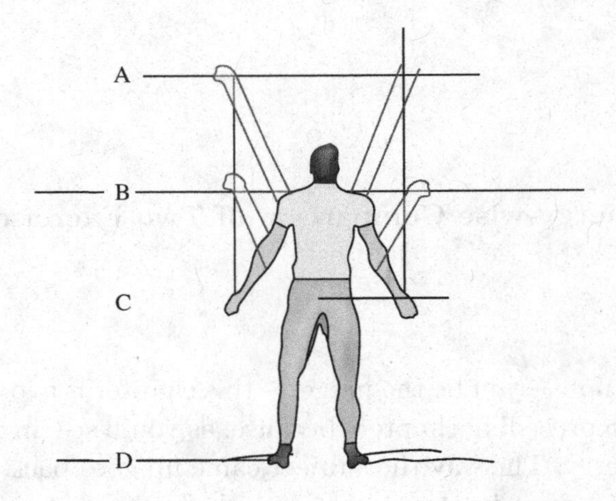

Fig. 9.1. Introducing Shad

To see what that all means, we need to postulate the existence of a person of interest, that is to say, an individual who can be counted on to perform ***both*** exercises. By way of introduction, here's the ***outline*** of the body of a rather athletic male (Fig. 9.1), whose actual identity and, in fact, more detailed description, are of no great consequence. With his identity thus immersed in the ***shadows*** we shall referred to him as ***Shad***, the implication being that a physically realizable Shad would actually have preferred being pictured with his ***back turned***, i.e. facing a direction ***into the page*** with the palms of his hands facing to the front so that his thumbs are extending outward. (Please ignore the extra lines; we intend to use them shortly!)

As things have, fortunately, turned out, we find our very accommodating subject in a gymnasium and agreeable to performing the weight-lifting procedure first, in which case he walks over to confront a ***barbell*** adorned with a large, round weight on each end and begins the exercise by bending forward and down, ***turning*** his hands so that the palms face **rearward**, fingers down, and grasping the bar in the ***conventional grip*** with the fingers of each hand curled around it, a grip he maintains throughout the exercise.

He then straightens up so that the bar is still oriented horizontally at vertical location **C** which, in body coordinates, is at the level of

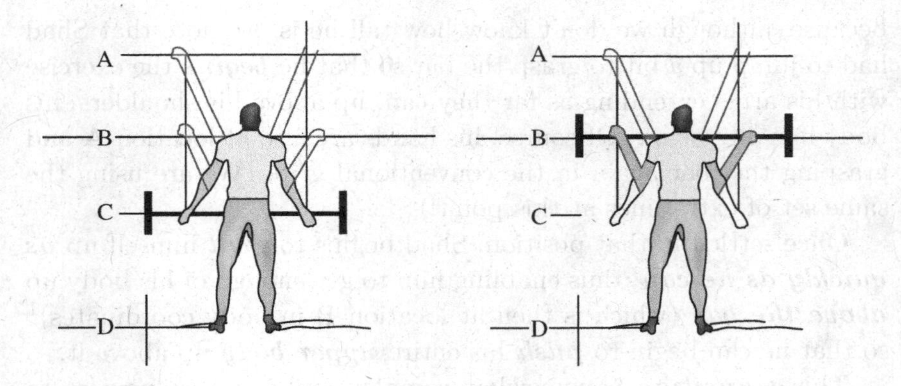

Fig. 9.2. Weight-Lifting Sequence

his hips (see Fig. 9.2) and, of course, in front of them. He then begins to pull the bar up as quickly as he can to location **B**, which is at the level of the top of his ***shoulders*** which begin to "hunch up", all the while ***turning his hands*** so that, at the end of this ***transition*** period, the palms face more or less ***upward,*** thus ***supporting*** the weight, while his ***arms*** have been ***flexed "akimbo"*** with the ***elbows*** pointing ***downward***. Actually, it has been found necessary to ***facilitate this transition*** by what is described as "***dropping down below*** the bar" which (having been "***cleaned***") remains at shoulder level in body coordinates.

All this in support of the next major action which is ***pushing*** the ("***cleaned***") bar ***up*** as fast as he can, in what one might describe as a "***jerking***" motion to location **A**, where his arms are extended ***upward*** as high as they can get. (Note the first use of extra lines in the figure!). There is also some leg work involved in that "dropping down" and straightening up activity but in the end we find it to add little to our story; basically, what has happened to the ***barbell*** is that it was first ***pulled up*** from **C** to **B** in body coordinates and then, similarly, ***pushed up*** from **B** to **A** in both body and ground coordinates, which at that point are identical.

Weightlifting over, our hero (carefully!) puts the weight back down on the floor and approaches the ***high-bar*** which is now located at what we now ***redefine*** as at a new level **A**. We need to do that because, although we don't know how tall he is, we note that Shad had to jump up a bit to grasp the bar so that he ***begins*** the exercise with his arms extending as far they can, up above his shoulders, his body hanging below. Of course, his hands are also at location **A** and grasping the, bar again in the conventional grip. (We are using the same set of extra lines at this point!)

Once settled in that position, Shad begins to ***Pull*** himself up ***as quickly as he can***, thus enabling him to get enough of his body up ***above the bar*** (which is then at location B in body coordinates)[1] so that he can begin to ***push*** his entire ***upper body*** up above it.

This ***transition*** is somewhat complex and is where novices to the exercise encounter the most difficulty. Strength-to-weight ratio is also important here because, for one thing, the faster one can ***pull up*** the easier it is to utilize remnant inertia, as in the weight lifting case, but performed up in the air! During transition Shad's arms also go through a transitory ***akimbo*** orientation, again with his shoulders hunched but this time with his ***elbows*** somewhat higher than his hands. At the end of this transition period, Shad has ***pushed himself*** up against the bar which ends up in position **C** across his hips, again in body coordinates. So, again, it's first ***pull up*** then ***push up***, this time against the pull of ***gravity*** on his body weight.

We don't really need an additional illustration for this procedure; it's essentially just Fig. 9.2 in reverse. The sequence is now from right to left, although the detailed "akimbonic" progression of the arms during transition might appear to differ in the two cases. Actually, what is occurring is that the two ***transition*** procedures are, in effect, ***inverse*** to each other! In the ***weight lifting*** procedure one tries to get in a good position to push up an external weight ***hung on a bar***, while in the ***muscle-up*** procedure we try to get in position to push our body up ***above a bar***.

The two ***procedures*** are thus seen to be almost identically ***inverse*** to each other and in summary to this point, let's ignore

---

[1]Which, we note, is just the opposite relationship to the bar as was required in the weightlifting exercise!

for the moment, how Shad gets the weight off the floor in the first case and his initial leap to catch the high bar in the second case.

The salient point in the ***first case*** is then that our hero lifts a ***dead weight*** through the vertical ***interval*** **C** to **A**. And in the ***second*** case he lifts ***his own*** weight through ***the same*** vertical interval although in the ***opposite labeling*** sequence. Of course, in terms of the ***physics*** of the situation, labeling doesn't matter; thus, if the two ***weights are equal***, Shad does the ***same*** amount of work in each case by lifting that weight up through the ***same vertical interval***.

So, that being the case, just how does it translate into ***physical*** terms? Indeed, what has Shad (unwittingly, of course!) accomplished in the process? Well, in physical terms, he has ***increased*** the ***potential energy*** of the ***weight*** against the ***force of gravity*** and by the ***same amount*** in each case! Which is to say, by an amount $P = WD$ where P is the incremental ***potential*** energy and W and D are (***whichever***) weight and the (***magnitude*** of the) ***vertical distance*** between locations **A** and **C**, respectively. Of course, when Shad puts the weight down in the first case and drops off the high bar in the second, all that potential energy is lost to the gravitational pull of the Earth; Sad but true!

Of course, there are the matters we previously "de-emphasised" — the difference between how high our hero has to jump in the second case, and how high level **C** is because he had to get the weight up off the floor to that level to start his exercise in the first case. Maybe we can talk about those details some other time!

In any event, that should suffice for simple, straightforward dynamics. However, we note, that, as a ***biological*** entity, Shad is subject, as are we all, to the laws of ***Thermodynamics*** and though clearly young and strong of body, his performance is not perfectly efficient. In other words, he must expend ***at least*** as much mechanical energy as expressed above in performing each of the required tasks but there must have been an increase in ***entropy*** in each case simply due to the mechanical ***inefficiency*** of his musculature of which there is a plethora! To begin with, there's your

biceps and ***triceps*** just to fold and unfold one's arms in pull or push activity.

However, perhaps of more importance are the "Lat's" (***Latissimus Dorsi***) on either side of the upper body; you can't really pull yourself up to the bar with any speed if your Lat's can't pull your arms down like they should. And, similarly, you can't push your way up off the high-bar if that ***Trapezius*** muscle from neck to shoulder is not doing its "super shrug" job. And we must not forget some smaller but very important forearm muscles that are required in order to change the orientation of your hands.

Now that we're on the subject, one quickly realizes that it is an endless task to track down all the contributions to the increase in ***entropy*** associated with the way in which Shad's muscles ***derive*** their ability to ***expend energy***. For example, one might think that the previous chapter might cover the increase of entropy involved in his exertions. There is, in fact a neat formula therein that expresses the increase in entropy of a ***mechanism*** that performs work in terms of the fractional increase in temperature the mechanism undergoes, the mechanism in this case being Shad's body. Unfortunately, as we know, a healthy body can foil potential temperature rises by such means as the generation of sweat. And then there is the entire phenomenology of breathing, pumping blood, etc. implying some familiarity with the field of ***Biomechanics***, quickly to be superseded by the ***Biochemical*** considerations associated with the conversion of nourishment into energy and the inefficiencies associated thereto. And on and on; I'm sure you've got the picture.

At this point, why don't we just agree that we ought to be satisfied in summing up the ***mechanics*** of what we have discussed thus far, thanking our shadowy friend for his contributions, and moving on to another chapter?

So, why don't we just?! Whereupon, we quickly realize that summing up is best conducted in terms of the basic nature of Shad's ***activities*** which, I hasten to assure you, are not the least bit Shady! And in that regard, I must emphasize that what we are ***really*** concerned with in particular, are the ***actions*** he applies to the ***bars of*** reference of which there are ***two***, the ***high bar*** in one exercise, and the ***Barbell*** bar, the one connecting the two ***weights***, in the other. And that action, we recognize, really comes down to being simply either a ***push or a pull***; Period! Note that, as far as the stress on Shad's muscularity is concerned, we're not much interested with the ***direction*** in which those actions are applied.

So, focusing on what ***happens to the bar*** in each case, rather than on Shad himself, allows us to view the two exercises from the same perspective. Consequently, the summing-up process turns out to be "summable" in terms of the diagrammatical sequences discussed above, but now, with the two exercises considered together, side-by-side, in effect as two vectors in close proximity, viz:

$$\begin{bmatrix} \text{(Bar-} & \text{(High-} \\ \text{Bell)} & \text{Bar)} \end{bmatrix} \Rightarrow \begin{Bmatrix} \text{[Push]} & \text{[Pull]} \\ \text{[Pull]} & \text{[Push]} \end{Bmatrix} \Rightarrow \begin{Bmatrix} [+] & [-] \\ [-] & [+] \end{Bmatrix} \Rightarrow \text{Matrix } m.$$

The ***column*** on the left in each diagram thus refers to the ***first*** activity, ***weightlifting***, and that on the right to the second activity, ***muscling up***. As we see, there are, therefore, ***four sectors*** of activity: as we recall, in ***weightlifting***, the ***bar*** is first ***pulled up*** from a lowest position and then ***pushed up*** to the highest, while in ***muscling-up***, we must view the ***bar*** as being first ***pulled*** down from its highest position and then ***pushed*** down to its lowest, all in ***body coordinates***, which, direction notwithstanding, is just the ***inverse*** of the weightlifting activity! In effect, in the second case, we might even consider Shad to be ***working against*** what amounts to ***Antigravity***, a putative force that tries to ***pull*** things ***up!***

We can then, arbitrarily, assign a ***plus*** sign to "Push" and a ***minus*** sign to "Pull" in either case, whereupon, lo and behold; our ***signature*** of (Matrix m) ***Complementarity*** has emerged once more! Most gratifying! And, if the goal of the combined pair of procedures is to achieve a balanced, incremental bit of exercise, the two constitutes a ***complementary pair*** since in each of the intervals we looked at, Shad finds a complementary ***push*** in one sector for each ***pull*** in the other and vice versa! Long may such complementarity prevail!

Finally, we have gained something from this exercise in investigating what began as a comment or two in a casual conversation

about exercise, something has emerged that we have not really stressed very much before but that turns out to be the ***essence*** of ***Complementarity*** at least as formalized by Matrix m. In essence that matrix is characterized by ***two binary vectors***, sitting side-by-side, and each composed of ***two independent components***, in this particular case, a ***push*** and a ***pull***.

But the unique feature of the matrix is the mutual ***inversion*** of the two vectors as a result of which, we emphasize again, that Matrix m is a ***symmetric*** matrix and its ***own transpose***; something we have also seen emerging graphically in our ***detailed comparison*** of the two exercises, ***including*** the associated ***transition*** procedures relative to the bar in the two cases!

So, here we go, on to Chap. 10 and Maxwell's equations, etc., but before we do so, I must thank Tom Manchester for reminding me that there is more to life than just pulling oneself over a high bar! Had he not done so, the next chapter would be Chap. 9!

10

Maxwell's Equations and the Electromagnetic Field

There is a long historical background associated with what has come to be called Electromagnetism. Electric and magnetic phenomena were well-known separately long before but the 18th and 19th centuries saw a greatly increased activity in both experimental and theoretical activity in both areas. The names ***Franklin***, ***Volta***, ***Coulomb***, ***Poisson*** and ***Gauss*** are prominent in accounts of the period (Krider 2006, APS 2006a, APS 2016a). Then in 1820 Hans Christian ***Oersted*** (APS 2008) discovered that an electric current can deflect a compass needle and shortly thereafter Andre ***Ampere*** showed that two parallel electrically conducting wires would attract or repel each other depending upon whether the associated currents were in the same or opposite directions (NIST 2018). And finally, Michael ***Faraday*** (APS 2001) showed that a changing current in one circuit would induce a current in a neighboring circuit as a result of the transient magnetic flux induced thereby and thus showing the way to the development of transformers, electric motors and generators.

The name of James Clerk ***Maxwell*** (ETHW 2021) is known nowadays as one of the most important in the history of physics (in those days he and others in his line of work were known as "Natural Philosophers" rather than physicists) and not just for his compilation of the set of equations associated with his name; he was acclaimed for his productivity in numerous areas. Maxwell was very much impressed with Michael Faraday's way of representing the strength of electric and magnetic influence in terms of "lines of force" and he sought to put something together that would highlight that concept.

Working with partial differential equations, the favored formalism of the day especially for the analysis of fluid phenomenology, it took him several years before he was satisfied with the outcome, a set of equations published in 1861 and 1862 (ETHW 2019a). But it was a momentous advance in the history of physics, not only because it united electricity and magnetism but because it also showed that ***light*** was an ***electromagnetic*** phenomenon which propagated with a speed that matched the experimentally measured values.

Soon Heinrich ***Hertz*** (ETHW 2019b) validated Maxwell's results with a pioneering series of experiments demonstrating the optical behavior — focusing, diffraction, etc. — of ***electromagnetic waves***, thus, inferentially verifying Maxwell's identification of light as an electromagnetic phenomenon. Also, it was not long before both Hertz and Oliver ***Heaviside*** (ETHW 2019a) were able to express Maxwell's equations in what is essentially the more easily manipulated vector notation. Maxwell himself would most assuredly have employed that formalism had it been available.

Even so, all is still not smooth sailing in expressing the equations; although there are only four of them, there are, of course, also four basic variables, two related to electricity (and to each other) and two to magnetism (also so related) plus an electric charge density and an electric current density. There seems to be no universally favored set of four and I wouldn't try to compute the various combinations and permutations available if one were to mix and match. Not to mention the choice of units. A quick tour through the Internet has gleaned a set that's O.K. and here it is:

$$\begin{aligned} \nabla \cdot \overrightarrow{D} &= \rho \\ \nabla \cdot \overrightarrow{B} &= 0 \\ \nabla \times \overrightarrow{E} &= -\frac{d\overrightarrow{\mathrm{B}}}{dt} \\ \nabla \times \overrightarrow{H} &= \frac{d\overrightarrow{D}}{dt} + \overrightarrow{\mathrm{J}} \end{aligned} \qquad (10\text{-}1)$$

The first two equations express Gauss's law for volume electric and magnetic ***charge density*** in terms of medium-dependent electric and magnetic field "***displacement***" variables, respectively, the zero on the RHS of the second equation reflecting the fact that, ***no magnetic charges*** had shown up in any experimentation (nor have they yet!). The third equation expresses, as the ***curl*** of a field vector, Faraday's discovery that an ***electric field*** can be experienced as circulating around a ***changing magnetic*** field and, similarly, the fourth equation expresses ***Ampere's*** discovery that a magnetic field encircles an electric current and Maxwell's realization that a changing electric field generates a magnetic field in the same way that an electric current does. The "***displacement***" vectors are given by $D = \varepsilon E$ and $B = \mu H$ with ε and μ being the free-space electric ***permittivity*** and magnetic ***permeability***, respectively, the inverted delta (known as a "Nabla") signifies the ***gradient*** operator, whereupon the dot product and cross product signify the divergence and curl, respectively of a field vector.

I would be remiss if I did not call to your attention the fact that what we see here is what would be a symmetrical, ***complementary*** combination of electrical and magnetic fields except for the presence of a symmetry-breaking term in that last equation added by Maxwell to express the encirclement of an electric current in a conductor by a magnetic field (as evidenced by the deflection of a compass needle). And, of course, the lack of a magnetic density term in the second equation.

There are also a couple of equations that relate our two fields to ***potential*** functions (see Chap. 12, "Noether's Theorem" and "Gauge Theory")

$$\vec{B} = \nabla \times \vec{A} \text{ and}$$
$$\vec{E} = -\nabla V - \frac{d\vec{A}}{dt} \tag{10-2}$$

where $\vec{A}$ and V are (so far) ***arbitrary*** vector and scalar potential functions, respectively. Arbitrariness being the operative word, there should be no objection to attaching another function to each as follows

$$\vec{A} \Rightarrow \vec{A} + \nabla\alpha$$
$$V \Rightarrow V - \frac{d\alpha}{dt}. \tag{10-3}$$

Inserting the thus-transformed terms into Eq. (10-2), and lo and behold — we find them to emerge once more unchanged!

$$\begin{aligned}\vec{B} &\Rightarrow \nabla \times \vec{A} + \nabla \times \nabla\alpha = \nabla \times \vec{A} \\ \vec{E} &\Rightarrow -\nabla V + \frac{d\nabla\alpha}{dt} - \frac{d\vec{A}}{dt} - \frac{d\nabla\alpha}{dt} = -\nabla V - \frac{d\vec{A}}{dt}.\end{aligned} \tag{10-4}$$

In other words, the relationships expressed by Eq. (10-2) are ***invariant*** to the given transformations which, ***together***, constitute what is known as a ***Gauge*** Transformation in, we are obliged to note, another example of ***complementarity***!

Incidentally, it is of interest to combine the vector and scalar potential functions into a four vector

$$A^{\mu} = (V; \vec{A}), \tag{10-5}$$

whereupon, the transformation can be written in the form $A^{\mu} - d^{\mu}\alpha$, known in the trade as ***covariant derivatives***. We shall have more to say about that kind of transformation when we talk about Gauge theory in Chap. 13.

Looking over what we have written down so far, it's also rather obvious that Maxwell's equations basically feature four ***complementary*** entities, two fields. E and M (or B and D) and two operators ∇. and ∇x, the so-called "dot" (or "scalar") product and the "cross" (or "vector") product, respectively, which is reminiscent of our set of four fundamental fermions or the four nucleotides basic to DNA.

In the meantime, note that we have not yet addressed what is probably the major achievement of Maxwell's equations, namely the inclusion of ***light*** under the taxonomical rubric of ***electromagnetism*** by showing that electromagnetic fields propagate as waveforms with a speed identical (at least in free space) to that of light! What we require in that regard is an easily derived wave equation, so we now consider a medium completely free of currents or sources, either electrical or magnetic. (Of course, as per Maxwell, there are no magnetic sources to begin with.) The benefit here is that

Maxwell's equations thereby reduce to two, in fact simply Faraday's and Ampere's laws,

$$\nabla \times \vec{E} = -\frac{d\vec{B}}{dt}$$
$$\nabla \times \vec{H} = \frac{d\vec{D}}{dt} \tag{10-6}$$

or, using the field equivalents of B and D

$$\nabla \times \vec{E} = -\mu \frac{d\vec{H}}{dt}$$
$$\nabla \times \vec{H} = \varepsilon \frac{d\vec{E}}{dt}. \tag{10-7}$$

Which, we observe are a (not quite perfectly symmetrical) ***complementary*** pair of equations. Then, taking the curl of the upper equation we get, (using the second equation)

$$\nabla \times \nabla \times \vec{E} = -\mu \nabla \times \frac{d\vec{H}}{dt} = -\mu d \frac{(\nabla \times \overline{H})}{dt} = -\mu\varepsilon \frac{d^2\vec{E}}{dt^2} \tag{10-8}$$

and, at the same time, we see that the LHS becomes

$$\nabla \times \nabla \times \vec{E} = \nabla \left(\nabla \cdot \vec{E} \right) - \nabla^2 \vec{E} = -\nabla^2 \vec{E} \tag{10-9}$$

where we have invoked the source-free assumption again. Combining these two results we then have what constitutes the required wave equation

$$\nabla^2 \vec{E} = \mu\varepsilon \frac{d^2\vec{E}}{dt^2}, \tag{10-10}$$

where, it is found, that $\sqrt{1/\mu\varepsilon} = c$, the speed of light in vacuum, all quantities having been measured and confirmed in various ways through the years. And, of course, repeating the procedure beginning with the lower of Eq. (10-7) produces Eq. (10-10) but with H substituting for E and recalling that $\nabla \cdot \vec{H} = 0$ as per Maxwell.

Note that there has been no mention anywhere in the above of a ***preferred reference system*** or, indeed, of ***any*** reference system at all. Or of what happens to Maxwell's Equations in going from one reference system to another. The implications of this state of affairs are discussed in the next chapter when we consider the Theory of Special Relativity.

11

Spacetime and The Ethereal Road to Relativity

Nowadays a lot of people are giving a lot of thought to the nature of space and time. It's really nothing new; so did the philosophers of the classical ancient world and in fact, in the 18th and 19th centuries that saw an explosion of both theoretical and experimental activity. Most influential in that regard was Isaac Newton who regarded "absolute" space as remaining "immutable" without "reference to any external object" and "absolute time" as flowing uniformly and similarly constrained (Newton 1729).

The transverse wave theory of light was established in 1818 by ***Fresnel*** (APS 2016b, Tietz 2018) who subscribed to the Newtonian philosophy but envisioned Newton's "absolute space" to be filled by a fluid, the "***ether***". The word itself was not new; according to the dictionary "Ether" is "An imaginary substance regarded by the ancients as filling all space beyond the sphere of the moon and making up the stars and planets", a somewhat "ethereal notion". But for Fresnel, the substance was real with optical properties that were the basis for his dynamical theory of optical phenomena. At the same time there were problems with the notion; as a simple example, consider the propagation of a transverse wave in an elastic medium, plane polarized in the y direction and moving in the x direction. The wave equation is

$$dy^2/dx^2 = (\rho/k)\,(dy^2/dt^2) \tag{11-1}$$

where ρ and k are the planar mass density of the medium and the transverse restoring force per unit displacement in y, respectively. We then find the velocity of propagation of the displacement to be

$v = \sqrt{k/\rho}$ which implies that the apparently large velocity of light in space would require a very large restoring force, a very small density or both. Other problems arise including the need for different values for density and restoring force as a function of wavelength, the fact that no resistance to astronomical bodies had ever been detected, the lack of models for the intermingling of ether and the microscopic arrangement of ponderable matter all adding up to a rather daunting state of affairs to confront in a scientific manner.

However, a major issue whose resolution felt amenable to experimentation was whether the ether is actually carried along by the movement of the local medium, either completely, partially or not at all. In 1951, I wrote an MS thesis paper entitled "The Experimental Basis of Special Relativity". (A sociological aside: By coincidence, I was then about the same age as Albert ***Einstein*** when he published his groundbreaking theory! However, he had not spent several years in the military as I had, instead honing his vastly superior intellect on problems of cosmic import. And, since the theory had already been developed by 1951, there was of course little left for me to accomplish! Or so it seemed at the time.)

At any rate it turns out that there was indeed quite an interesting experimental background and much of it was, in fact, related to the above issue. Some of the experiments were optical in nature and some electromagnetic. Also, it can be shown that they can be further organized in terms of a single parameter, the ratio $r = v/c$, where v is the ether velocity and c is the free-space velocity of light and, in fact, either r itself or squared, otherwise referred to as ***first-order*** or ***second-order*** experiments.

In the thesis, I looked at a total of nine experiments, six first-order and three second-order. Three of the first-order experiments were optical in nature and three were electromagnetic. Of the second-order experiments, one was the famous (and definitively null) ***Michelson-Morley*** attempt to measure the motion of the Earth through the ether using a very precise optical interferometer (APS 2007), but conducted before Einstein's theory was published, and the other two were electromagnetic and conducted afterward to see if the theory was experimentally verifiable. Which, of course it was.

The very first experiment was really carried out by ***Fizeau*** just to show that the measured velocity of light was indeed modified as it traversed a moving ponderable medium and to a predictable extent (APS 2010). The medium in this case was water flowing through the same cylindrical cavity as traversed by the light and the result was agreeably positive lending encouragement to further experimentation to detect the actual presence of that mysterious medium — the ether. The first-order experiments were then carried out with great care, more sophisticated design and high hopes but to no avail; the ether declined to show its presence. At that point, the highly respected physicist, ***H.A. Lorentz*** introduced his celebrated ***electron theory*** (Wilczek 2012, Lorentz 1909, 1902), a mechanistic model that visualized the electron as a sphere of electric charge density, deformable by motion through the ether as a result of internal ***electrodynamic*** forces.

What the theory showed is that the deformation occurred in such a way that no ***first-order*** experiments ***can*** detect its cause by ethereal drag even if such there be! On the other hand, ***second-order*** effects were unquestionably predicted to be detectable so when, as it turned out, they were not, consternation ensued and it was back to the drawing board. What emerged was another mechanistic explanation known at the time as the ***Fitzgerald-Lorentz*** contraction (Tatum 2020), a ***transformation*** that involved ***both*** space and time coordinates as a function of ether velocity relative to the free space velocity of light but in a more complicated way that Lorentz arrived at as a result of a more sophisticated look at his electron theory.

Here's what the transformation amounts to: assuming a constant velocity, ν, of a body moving along the x direction associated with the coordinate system of a stationary observer, distance, x', and time, t', as experienced by an observer in the ***moving*** system are given in terms of the corresponding measurements by the stationary observer as

$$x' = \frac{x - vt}{\sqrt{1 - (v/c)^2}}$$
$$t' = \frac{t - vx/c^2}{\sqrt{1 - (v/c)^2}} \tag{11-2}$$

which can be immediately simplified as

$$\begin{aligned} x' &= \frac{x - \gamma\tau}{\beta} \\ \tau' &= \frac{\tau - \gamma x}{\beta} \end{aligned} \tag{11-3}$$

by setting

$$\begin{aligned} ct &= \tau \\ v/c &= \gamma \\ \sqrt{1-\gamma^2} &= \beta. \end{aligned} \tag{11-4}$$

In terms of the main theme of this book, either of these versions expresses a ***complementary*** combination of spatial and temporal variables. It may be a bit easier to see that if we simplify the second expression above a bit more by setting

$$\begin{aligned} x &= X \\ \tau &= T \\ \beta x' &= X' \\ \beta\tau' &= T' \end{aligned} \tag{11-5}$$

to obtain

$$\begin{aligned} X' &= X - \gamma T \\ T' &= T - \gamma X \end{aligned} \tag{11-6}$$

or in vector/matrix notation

$$\begin{pmatrix} X' \\ T' \end{pmatrix} = \begin{pmatrix} 1 & -\gamma \\ -\gamma & 1 \end{pmatrix} \begin{bmatrix} X \\ T \end{bmatrix}. \tag{11-7}$$

And we note, the matrix on the RHS is very much like the algebraic characterization we found for both DNA and the four basic elementary particles of our Alternative Model of the elementary particles as, namely,

$$m = \begin{pmatrix} 1 & -1 \\ -1 & 1 \end{pmatrix}, \tag{11-8}$$

our ***complementarity signature*** matrix as in Chap. 5. The resemblance is an equality, (i.e $\gamma = 1$) only for photons.

So much for the ***complementarity*** of what has become known as the Lorentz transformation (somehow contribution has tended to disappear in the literature). But what does it have to do with relativity? The answer is "just about ***everything***". In the first place, the concept of the ether was still omnipresent in the physics community; almost everyone still thought of the Lorentz transformation as expressing a mechanistic way of compensating for ethereal drag, including, of course, Lorentz himself. Everyone, that is, except Albert Einstein as attested to by the publication in 1905 of his first paper on the ***Theory of Relativity***, the one entitled "***On the electrodynamics of Moving Bodies***" (Einstein 1905).

Albert had a lot of faith in experimentation and his conclusion thereto was quite sensible, namely that it makes no sense to postulate the existence of an ether if we can't detect it; so let's just forget about it![1] Furthermore, he showed that the measurements made by various inertial observers are explained ***if*** and ***only if*** one ***transforms*** them according to the Lorentz Transform. Which, implicitly, assumes a constant value for the speed of light (or any electromagnetic radiation) in empty space. In fact, since all measurements of that speed get the same answer, let's call it an ***Invariant***! All of which jibes with what Maxwell's equations have to say about the effect of varying inertial measuring systems on the speed of light — ***nothing***! Light travels in space at the same rate no matter who emits or receives it.

In summary, the Theory of (Special) Relativity says that all inertial systems are seen to be ***equivalent*** as long as we transform the spatial and temporal measurements they make according to the

[1] Note the Alternative Model's philosophy regarding Quarks!

Lorentz transformation. The laws of physics are the same and have the same form in each and every one of such systems.

In 1946, Einstein published a book with the title "The meaning of Relativity" (Einstein 1946). It's a small book — 135 pages including the Index — but it contains a lot of insightful explanatory material not only of Special Relativity but of its big brother, the Theory of General Relativity, which we'll get to shortly. After all he wrote from the vantage point of the forty years, he'd had to ruminate about it (only thirty years for the general theory). And here's what he said in the first sentence of the book: "*The theory of relativity is intimately connected to the theory of space and time.*" And somewhat later: "*it is neither the point in space, nor the instant of time, at which something happens that has physical reality, but only the* ***event*** *itself. There is no absolute (independent of reference) relation in space, and no absolute relation in time between two events, but there is an absolute (independent of the space of reference) relation in space* ***and*** *time.*" (My emphasis).

From which we can infer the existence of an ***invariant***, in fact one that has been described as characterizing the Theory of Relativity and is known as the "***Proper time interval***". Actually, the existence of such an entity can be viewed as a ***consequence*** of the ***Lorenz Transformation***, which emphasizes the importance of the latter to the theory. Consider two closely spaced events and the associated increments in ***both*** space and time. Using the nomenclature in capital letters as per the above, the relevant Lorentz expressions are then

$$\begin{aligned} \Delta X' &= \Delta X - \gamma\Delta T \\ \Delta T' &= \Delta T - \gamma\Delta X. \end{aligned} \tag{11-9}$$

Squaring both expressions and subtracting we obtain

$$\left[(\Delta T')^2 - (\Delta X')^2\right] = (1-\gamma^2)\left[(\Delta T')^2 - (\Delta X')^2\right] \tag{11-10}$$

which, upon reverting to the original nomenclature then produces

$$(c\Delta t')^2 - (\Delta x')^2 = (c\Delta t)^2 - (\Delta x)^2.$$

Thereby expressing the ***invariance*** of the *difference between the square of the time increment and the space increment*, a radical departure from the Newtonian immutable views of space and time!

At this point in our discussion we note two things: one is the change in how space and time are to be viewed in the post-Relativity world. Again in Einstein's words: " — the laws of nature will assume a form which is logically most satisfactory when expressed in the space-time continuum. Upon this depends the great advance in method which the theory of relativity owes to Minkowski. Considered from this standpoint, we must regard x_1, x_2, x_3, t as the four coordinates of an ***event*** in the four-dimensional continuum (Einstein 1946)." (My emphasis). To which we might append the historical note that it actually took Einstein a while to achieve a belated recognition of Hermann Minkowski's radical 1907 introduction of that four-dimensional continuum as the necessary venue within which relativistic geometry exists.

In his own words, here's what Minkowski had to say in a speech he gave in 1908: "Henceforth, space by itself and time by itself, are doomed to fade away into mere shadows, and only a kind of union of the two will preserve an independent reality (Minkowski 1908)." Of course, from the vantage point of his later view of spacetime and the central importance of the Lorentz transformation thereto, Minkowski tended to denigrate Einstein's contribution a bit! He shouldn't have but then he had also managed to derive Special Relativity himself but solely based on geometrical transformations in his four-dimensional Spacetime!

The second noteworthy thing is that the above invariance is also an expression of ***complementarity***! That is to say we now have an invariant quantity, say $(dS)^2 = (cdt)^2 - (dx)^2$ that owes its existence to the ***complementary*** contributions of ***both*** spatial and temporal quantities. In other words, the theory of ***relativity*** is another ***manifestation*** of the ***Principle of Complementarity!*** At least, that is, as far as the ***inertial frames*** of the theory considered heretofore are concerned; we have not yet mentioned the more general framework of the General Theory of Relativity. So, let's talk about it!

12

Noether's Theorem and Gauge Theory

Don't be alarmed; this chapter is not a diversion; it is indeed pertinent to our main story line! The previous section featured a comparison of DNA and the basic ideas of the Alternative Model; that was what, as documented in the previous book, initiated my original focus on ***complementarity***, a concept whose generalization as a principle promised to have fundamental significance. In contrast, here we begin with two subjects that are ***already*** so regarded, beginning with ***Noether's Theorem***, which has been described as forming "an ***organizing*** principle for all of Physics" (Neuenscwhander 2011). The theorem pairs the ***invariance*** of an entity governed by a ***symmetry*** principle to changes of reference with the ***conservation*** of an associated dynamic entity ***within*** a ***given*** reference system. More explicitly, each member of the pair ***implies*** the other: finding an entity that is invariant to ***change in reference*** implicates the existence of an associated entity whose value never changes as measured ***within*** a particular, associated reference. ***Conversely***, awareness of an entity whose value is constant within a given reference implies that there exists an associate entity invariant to ***change*** of reference.

But consider the larger significance: that ***pairing*** — invariance and conservation — clearly constitutes a ***manifestation*** of ***Complementarity***! Thus, in epistemological terms, the acknowledged ***elementarity*** of Noether's Theorem suggests that we recognize ***Complementarity*** as the very ***essence*** of ***Elementarity***; In other words, if something is known to be ***elementary*** — truly elementary — then it must be part of a ***complementary*** pair. And, if we discover such an elementary entity we can expect its ***complementary partner*** to be lurking in the woods just waiting to be discovered as well! In summary, Noether's Theorem may itself be viewed as encompassed within what might be termed a ***Principle of Complementarity***! Which of course is really what this book is all about: to validate the universal applicability of that principal.

In this chapter, we shall take a closer look at what ***Noether's Theorem*** is all about and in that regard to the ***historical*** origin of the theorem is germane. Our review of that history draws upon the book by Neuenschwander (2011).

As is well known, the theorem was developed by Emmy Noether, a remarkably influential mathematician of the early middle part of the 20th century who overcame institutionally solidified gender bias to originate important mathematics primarily in modern abstract algebra, her main field of interest.

However, her development of the theorem (actually theorems: what she came up with was published in two parts each associated with somewhat different "ground rules") turns out to involve mainly a combination of ***differential geometry*** and the fundamental ***Hamiltonian*** mechanics of the end of the 19th century that we talked about to some extent in Chap. 7. It grew out of her efforts to solve a ***puzzling aspect*** of the Theory of ***General Relativity*** that emerged in a series of lectures presented in early 1915 by Albert Einstein to the assembled principal mathematicians of the Göttingen school headed by the illustrious David ***Hilbert*** in early 1915.

This was before Einstein had formulated the theory in its final definitive form. However, as a result of that interaction, both he and Hilbert then proceeded to accomplish that task shortly thereafter (Another well-known story but one we must forego relating herein!) and Hilbert, a champion of Noether and her intellect and abilities, subsequently asked her to attend a talk he gave on his approach to General Relativity. The puzzle referred to was stressed as part of the presentation; it had to do with an apparent lack of ***local*** energy ***conservation*** in the theory and appeared to be fundamental and unavoidable. Hilbert explicitly asked Noether to clarify the situation and her Theorem emerged in the course of doing so. Her explanation showed that Hilbert was ***correct***; the ***energy situation*** was indeed ***inherent*** in the ***theory*** and the explanation as to why that was the case was in fact a ***consequence of the Theorem.***

We shall have more to say about the failure of local energy conservation in General Relativity below. In Noether's Theorem it is described in terms of the ***symmetry*** groups that apply in the case of regions containing gravitational effects as contrasted with those that do not and (not to worry) by no means does it invalidate the theory of General Relativity. However, it turns out that her theorem itself is impacted in a radical way by Noether's use of it to clarify Hilbert's puzzlement; basically the theorem turns out to be ***inapplicable*** in those regions containing ***gravitational effects***. In fact, as we shall see below, the inapplicability is more general than that and it has to do with the ***local*** geometry of spacetime.

Now there are a variety of ways to approach the derivation of her Theorem; Noether happened to follow Hilbert in taking the Hamilton-Jacobi approach in seeking the extremum of an Action integral, that is the Integral of an appropriate Lagrangian, and setting the resulting Euler/Lagrange equations to zero (see Chap. 7 for the use of such an approach in discussing the physics of our version of the elementary particles). However, in whatever approach is taken there is a similar requirement for the existence of ***invariance***, or, as per the above, a ***symmetry principle*** and that's where ***Gauge Theory*** comes in to save the day.

As is well known, in Gauge Theory the ***global*** invariance associated with a group of transformations can be assumed to apply on a ***local*** basis as well, ***provided*** that, as stressed in the admirable book by ***O'Raifertaigh*** on the origins of GT (O'Raifeartaigh 1997), the partial derivatives applicable to phenomena taking place in ***ordinary*** spacetime must be replaced in more complex (but by no means necessarily unusual) circumstances by what are known as ***Covariant*** derivatives. Basically, what that means operationally is that the calculus we are taught in, as in my case, first-year University, is no longer applicable in the more general situation. Instead of the familiar partial derivative $d_\mu(x)$, we must use an augmented formalism known as the ***covariant*** derivative

$$D_\mu d_\mu x = d_\mu + A_\mu(x). \tag{12-1}$$

Actually, the A_μ were long known to the mathematical community as ***connections*** and are ubiquitous in General Relativistic literature where they are known as ***Riemann-Christoffel*** connections. (Again, see Chap. 13 for use in regard to our elementary particles.)

After a tentative start in the first half of the 20th century, Gauge Theory has become the standard way to approach most of the modern physical world but the nomenclature therein is somewhat different; the mathematical "***connections***" are known to physicists as Gauge ***Potentials*** and express the fields associated with the phenomena of concern whether electromagnetic, gravitational or those associated with the elementary particles of the Standard Model, etc. The transformations referred to above are known as ***Lie groups*** or, more specifically in Gauge theory, as ***gauge groups*** whose algebra governs the values taken by the fields.

As we see from Noether's explanation, clearly, our concern here is with ***geometry*** and, in particular to begin with, the geometry associated with ***General Relativity*** which means the effects of gravitation which, in turn, means the ***curvature*** of ***Spacetime***. That ***General Relativity*** is indeed a ***Gauge Theory*** discovered by ***Ryoyu Utiyama*** in Japan early in 1954 (O'Raifeartaigh 1997). He did not share the Nobel Prize with Yang and Mills for the initial development of Gauge theory due to an unfortunate sequence of events regarding publication. Later, in the course of reflectively explaining that situation Utiyama goes on to say "that fields carrying a fundamental force — either gravity or electromagnetism — must in fact be those termed *connections* in mathematics, which are now called *gauge potentials.* That the concept of connections is indispensable in establishing a theory of interactions was the basic assertion of my paper (O'Raifeartaigh 1997)."

The "***gauge potentials***" to which Utiyama referred are expressed by the letter "A" in the equation for the covariant derivative shown above and, as we shall see, it is the key to describing the presence of ***curvature***. In the case of ***electrodynamics***, for example, A is also known as the "***vector potential***" whose derivative in Maxwell's equations is the magnetic vector ***force*** felt by a charged particle, for a time regarded as an artifice needed just to make those equations have predictive power and should feel a force only where a magnetic field actually exists.

However, in a "landmark" paper ***Aharonov*** and ***Bohm*** (1959) devised a famous experiment in which, they predicted, two identically charged particles traversing in a region ***outside***, but in the ***vicinity*** of another region within which the actual magnetic ***field*** was ***confined*** would, when they ***converge***, interfere in a way that depends on the particular vector ***potentials*** they experience in that traverse. The presence of what we can also describe as a "gauge potential" ***outside*** the actual region of confinement was predicted on a ***quantum mechanical*** basis, meaning that there is a ***phase*** associated with the ***region traversed*** by the particles (but not with the particles themselves) and it emerges ***only*** in the ***quantum mechanical*** approach because Quantum Mechanics is formulated in terms of ***complex algebra***. The happy ending to the story is of course that the experiment was indeed performed and with the expected result from which it was concluded that the electromagnetic potential is real and that ***electromagnetism*** can be viewed as a ***gauge*** phenomenon.

There is more on that subject later on but we have diverged a bit so back to Hilbert's problem and Noether's efforts thereto: as Byers (and Utiyama) point out, since ***General Relativity*** is a gauge theory, the ***symmetry*** associated with it is that of a ***Lie group*** and in particular the group of "all continuous coordinate transformations with continuous derivatives" otherwise referred to as the "group of general coordinate ***transformations***" for which the symmetry associated with ***Special Relativity*** is a subgroup also known as the ***Poincare Group*** (Byers 1999).

That's quite a mouthful but what it comes down to is that the difference is really due to the underlying ***topology*** in the two cases: the general theory (as a theory of gravity!) must include the effects of ***gravitation*** which implies ***curvature*** of spacetime whereas Special Relativity ***does not.*** In the ***Special case,*** the "energy-momentum tensor" of the theory is "divergence free" and energy is conserved on a local basis. What Noether showed is that in the ***General case*** "it has ***no meaning*** to speak of a ***definite localization*** of energy" because the divergence freedom of the energy momentum tensor is "***gauge dependent***: i.e. it is not covariant under general energy transformations" and there is thus no "meaningful law of energy conservation". In other words, Noether's Theorem does ***not apply*** to ***General Relativity*** where gravitational effects predicted by that theory obtain — in essence, where there is ***Curvature*** (O'Raifeartaigh 1997).

But it does apply in the case of ***electrodynamics*** where the gauge group is ***U(1)*** the group of simple ***rotations*** as in, for example, the encirclement of a cylinder. And, I should reemphasize, on a ***local basis*** in the case of the elementary particles of our ***Alternative Model***! Also, as per ***Yang*** and ***Mills*** in their Nobel prize winning paper, particle ***isospin*** can be discussed in terms of the ***two by two matrices*** of the gauge group ***SU(2)*** which is the cover group for the group ***SO(3)*** of rotations in three space. And, to reemphasize once again (!), also as ***manifested*** by the elementary particles of our Alternative model, wherein ***Isospin*** is explicitly associated with the inherent ***twist*** of our ***solitonic Moebius strips***, also governed by the gauge group ***SU(2)***!

Before we proceed to the next chapter there is something that seems apparent when we stand back and take a long look at Noether's Theorem once more, at least the Hamilton-Jacob y derivation thereof, whose key feature is obtaining an extremum of the ***Action***. As the analysis proceeds, ***Action*** never actually disappears and, in fact, at the end we are presented with a formalism that features it in the form of ***pairs*** of variables (the ***complementary pairs*** discussed above) whose dimensions ***must*** multiply to that of ***Action***, namely ml^2/t, in essence the product of mass and the time rate of change of an area.

In any event, it would appear that Noether's Theorem is an ***inevitability*** and that the pairs of quantities that manifest it are ***necessarily*** complementary. It may not have escaped you that in a way we have been talking about what amounts to "***unquantised***" Quantum Mechanics; the only thing missing is h, the quantum of action!

III

General Relativity and the Geometry of Spacetime

13

General Relativity

In the first place, whereas, in the preceding, discussion was primarily ***algebraic*** in nature (as it was for Einstein), the General Theory is intimately concerned with the essential ***geometry*** of Spacetime, an entity with ***Character*** and possessing ***attributes.*** You may recall the bit of "arcanity" presented in the chapter on ***Noether's theorem*** and ***Gauge theory***, to wit: "— as Utiyama and Byers point out, since ***General Relativity*** is a ***gauge*** theory, the symmetry associated with it is that of a ***Lie group*** and in particular the group of 'all continuous coordinate transformations with continuous derivatives' otherwise referred to as the 'group of general coordinate transformations' for which the symmetry associated with ***Special Relativity*** is a subgroup also known as the ***Poincare*** Group". It has been said that what Minkowski discovered was that Einstein's Special Relativity "— was nothing other than the theory of ***invariants*** of a ***definite group*** of linear transformations of R^4, namely, the ***Lorentz group***." (my emphasis). As I said above, a bit of a put down that Einstein apparently justifiably resented.

Anyhow, Lorentz or Poincare, the way I see it, what's important from a ***utilitarian*** point of view is the Lorentz ***transformation*** between ***inertial*** systems and the inferred invariance thereof (see above). It's something that's invoked all the time. Torretti (1983) also puts the subgroup characterization a bit differently, he speaks of the Minkowski discovery as a "local — tangential — approximation to General Relativity". What that means is just that although the Riemannian Space of General Relativity may exhibit various

distortions, e.g. curvature and (perhaps)[1] even torsion, a ***sufficiently small*** region of it may be considered "flat" enough for Minkowski's R^4 (or Einstein's Special relativity!) to apply. No big deal; it sounds like learning Calculus, doesn't it?

On the other hand, Einstein realized full well that his ***Special Relativity*** theory was in no way broad enough to encompass the kinematics and dynamics of real life, restricted as it was to inertial systems and flat space, Minkowski's spacetime notwithstanding. And he needed to be able to take ***acceleration and gravity*** into account, in regard to, of which it has been known for a long time that, what's known as "***inertial***" ***mass*** and "***gravitational***" ***mass*** are experimentally equal to high precision, ***gravitational*** mass being measured. For example in a balance against a calibrated reference, and ***inertial*** mass as what we observe as that of a hockey puck sliding along on the ice under the urging of an accelerating player. Or, a ***rocket*** reacting against the thrust of the fuel ejected out its back. Which is why ***all bodies***, no matter their weight fall with the same ***acceleration history*** (see Chap. 7).

And which leads us to Einstein's "***Principle of equivalence***". As related here and there, according to the good professor himself, he was looking out of a window, at some window washers on a tall building. Imagining what it would be like for one of them if he fell, he realized that the man would feel ***weightless***; ***in free-fall no force of gravity is felt*** even though it is producing an acceleration ***because*** Albert concluded, ***Gravitation and acceleration*** are ***equivalent manifestations*** of the same phenomena. Had the man been in a spaceship at a commensurate altitude (imagine a ***very*** tall building) with the proper forward velocity he would be in orbit which means that he is always really ***falling*** towards the center of the Earth but with a centrifugal acceleration commensurate to the orbital altitude. In both cases, the man feels no gravitational force and he is moving along what's known as a

[1]The chapter on the Alternative Model of the elementary particles is relevant here, particularly the Knot connection.

Geodesic, meaning a path that is always parallel to itself because there is nothing to cause the mover to move to a different path!

Noting that the planets in our solar system are in similar geodesics, each for its individual speed and that light itself is caused to bend its path when traversing a trajectory in the neighborhood of a star, Einstein came to a ***remarkable conclusion***; the reason small bodies in the vicinity of a large ponderable body assume curved trajectories is that the ***space*** in that vicinity is ***curved*** as a result of the large body's ***mass***: ***Gravitation is not really a force like the tension in a spring or that hockey player pushing the puck; It's a manifestation of the curvature in the vicinity of the large body due to its mass.*** And there has to be a way to equate that curvature to that mass! Eventually he realized that the mathematics with which he was familiar was inadequate to that task and he enlisted the help of his old friend from university days, the mathematician, Marcel ***Grossman*** (Janssen and Renn 2015).

It turned out that the required mathematics was the ***tensor calculus***, a formalism that's not usually taught to undergraduates (I believe; but it's been a very long time since I was an undergrad!) but it's really not as complicated as one might think. It involves a way to manipulate partial derivatives festooned with ***superscripts*** and ***subscripts*** in order to keep track of their purpose and role in the proceedings and I like to think of the ***Tensor*** Calculus (TC) as an ***algebra*** of such ***festoonery.*** Einstein gives a short treatise on the care and feeding of the TC in the small book I mentioned above as do many books on relativity. But don't be alarmed; I'm happy to tell you that we don't really need go that far in this book.

Nevertheless, in this book as in my previous book, we will be concerned to some extent with the ***formalism*** of General Relativity (GR) because we are interested in the differential ***geometry*** of a ***torus*** wherein the curved and twisted trajectory of a point moving along the (putative) associated torus ***knot*** evokes the notion of a ***connection*** to explicate the knot's topology. We talked about connections before in regard to Gauge theory and they are of course well-known as fundamental to the formulation of General Relativity which, as we mentioned, is known to be a Gauge theory.

In GR, the terms on the ***left-hand side*** of the world famous ***Einstein equation*** can be derived by the repeated ***contraction*** of R^{c}_{dab}, the Riemann-Christoffel ***Curvature Tensor*** (RCCT), a fundamental formalism which can be seen to emerge from a consideration of the **"parallel" transport** of a vector between two points along each of two ***different paths***. What that means is that the miniscule arrows we might visualize as describing the progress of each path precisely follow each other, "head to tail" so to speak ***because*** they are not subject to any influence that will cause them to deviate from that path.

The ***discrepancy*** that results between the two paths if they exist in non-Euclidean space is customarily formalized as the ***commutator***[2] of two covariant derivatives, each associated with its own path connecting the two points, viz:

$$(\nabla_a \nabla_b - \nabla_b \nabla_a) V^c = R^{c}_{\text{dab}} V^d + 2T^{e}_{ab} \nabla_e V^c \tag{13-1}$$

where V^c is the transported vector, the $\nabla_\alpha = d/d_\alpha + \Gamma^{e}_{c\alpha}$ are the ***covariant*** derivatives we discussed above and the $\Gamma^{e}_{c\alpha}$ are the ***connections***, in particular, the ***Christoffel*** Symbols (CS) of the second kind which have to be computed for each particular case (done in the previous book but not repeated here since we are not concerned with that detail).

In brief, the rightly celebrated ***Einstein Equation*** is

$$R_{\mu\nu} - \frac{1}{2} g_{\mu\nu} R = -8\pi G T_{\mu\nu}. \tag{13-2}$$

Here $R = g^{\mu k} R_{\mu k}$ is the ***Curvature Scalar*** (CvS), the $g^{\mu\kappa}$ are terms in the ***metric*** which, being specific to the nature of the curvature also have to be computed, we shall encounter in the next chapter because it's specific to the nature of the curvature, and

$$R_{\mu\kappa} = R^{\lambda}_{\mu\lambda\kappa} = (d\Gamma^{\lambda}_{\mu\lambda}/d\mathrm{x}^{\kappa} - d\Gamma^{\lambda}_{\mu\kappa}/d\mathrm{x}^{\lambda}) + (\Gamma^{\eta}_{\mu\lambda}\Gamma^{\lambda}_{\kappa\eta} - \Gamma^{\eta}_{\mu\kappa}\Gamma^{\lambda}_{\mu\eta}) \tag{13-3}$$

[2]Itself a manifestation of Complementarity!

is the ***Ricci Curvature*** Tensor (RCT), contracted from the Riemann-Christoffel Curvature Tensor (RCCT)

$$R^{\lambda}_{\mu\nu\kappa} = (d\Gamma^{\lambda}_{\mu\nu}/d\mathrm{x}^{k} - d\Gamma^{\lambda}_{\mu k}/d\mathrm{x}^{\nu}) + (\Gamma^{\eta}_{\mu\nu}\Gamma^{\lambda}_{k\eta} - \Gamma^{\eta}_{\mu k}\Gamma^{\lambda}_{\mu\eta}) \qquad (13\text{-}4)$$

and

$$T^{\mu}_{\mu} = T_{\mu v} g^{\mu v}, \qquad (13\text{-}5)$$

is the ***Energy Scalar***, contracted from $T_{\mu v}$, the ***Energy-Momentum*** Tensor also known as the Stress-Energy tensor. (As you might surmise, ***contraction*** means you equate a ***superscript*** and a ***subscript*** thereby reducing the amount of festoonization by two!)

In other words, the ***Einstein equation*** equates ***spatial curvature*** to local ***energy-momentum*** (stress-energy) with the key entity in its construction being which, in principle can be ***calculated*** by examining how vectors behave in traverse over closed circuits of manifolds in the spatial regions of interest as per the above.

It turns out, perhaps somewhat unexpectedly, that General Relativity is of more than academic interest to our Alternative Model of the elementary particles: the main purpose of the next chapter is to show how the basic elementary particles of our model exist as **Sine-Gordon *solitons*** and we invoke GR to supply the equivalence of ***curvature*** as per the ***left side*** of the Einstein equation to localized ***energy*** which, according to Einstein's famous $E = mc^2$, is another way to view ***mass***. So, on to the next chapter where all will (hopefully) become clear!

14

Differential Geometry and "The Action Enigma"; an Application

Although the ***usual scale*** of application of General Relativity is the ***cosmos***,[1] at some point in this chapter we shall employ it in so unorthodox a manner as to violate all rules of technical etiquette: we are going to use it to show how our Alternative Model (AM) ***particle model*** can arise out of ***spacetime*** and, in the process, to justify a statement we made earlier concerning the time-honored ***action integral***. I was going to request that you don't tell anyone but, recently, I came across a paper ***published, in 1919*** by Albert Einstein, no less, who investigated the ***possibility*** that ***gravitational*** effects might have some ***bearing*** on the nature of the ***elementary particles***. He concluded that "— there are ***reasons for thinking*** that the elementary forces that make up the atom are held together by ***gravitational*** forces (Einstein 1919)". Albert was actually thinking of the interior of the electron as some kind of localization of electromagnetic effects held together by ***gravitational forces***!!

A quick note before we proceed: you may recall the matter brought up in Chap. 12 concerning the integral of the quantity $K - V$, the difference between kinetic and potential energies, designated as the "***Action Integral***" (in what follows we will call it the ***Lagrangian***). The question was why the ***difference*** between the two terms appears and not the **sum** and some very good reasons were suggested in that Chapter. Nevertheless, here we re-examine the question in terms of our Alternative Model of the Elementary Particles. However, where

[1] Actually, our vaunted GPS system won't work unless GR effects are accounted for!

we were heretofore concerned primarily with the ***algebraic*** geometry of ***Flattened*** Moebius Strips (FMS) and their interactions, that is, with ***FMS*** as full-fledged elementary *particles*, at this point, we embark upon a different kind of investigation wherein our particles are presumed to have a life ***before*** they become quite ready to participate in that role. Thus, in this chapter we will be talking about the nature of these nascent particles mainly in terms of their ***differential geometry*** in one form or another.

Going back to my referenced book for a minute, as I said in the Introduction it started out as a way to summarize the work I had carried out over a period of some years and it makes reference to a short list of papers I had published in a knot theory journal. That's a valid venue because there is, in fact, a fundamental relationship between knots and Moebius Strips (MS). Knot Theory is of course a massive subject in its own right as a part of mathematical topology, but my work is concerned with only a miniscule part of that scope.

To quote the first paper of reference "Here, beginning with two rudimentary torus knots, the unknot and the trefoil knot, we develop a unique approach to understanding the elementary particles of physics in terms of a *visualizable* reduction of all particles — fermions and bosons, hadrons and leptons to a common topology". The paper goes on to characterize particles in the model "not as discrete, point-like objects in a vacuum but as *localized distortions in an otherwise featureless continuum* that supports torsion as well as curvature." In fact, it has been shown that our nascent elementary particles take the form, in that continuum, of what's known as ***Solitons***, specifically, as per their mathematical description, ***Sine-Gordon*** Solitons and the ***resolution*** of the ***Action problem*** is to be found in the associated mathematics. However, before we get that far, we need to talk about knots and their relationship to our nascent particles.

One way to illustrate the relationship is to cite the readily verifiable fact that the ***boundaries*** of the four basic FMS can be viewed as ***knots***, specifically, the folded-over unknot and the trefoil knot, as per the paper I quoted above. Actually, the boundary is representative of all similar paths on the MS. Thus, another way to look at it is to view the MS as a strip or ribbon consisting of a parallel

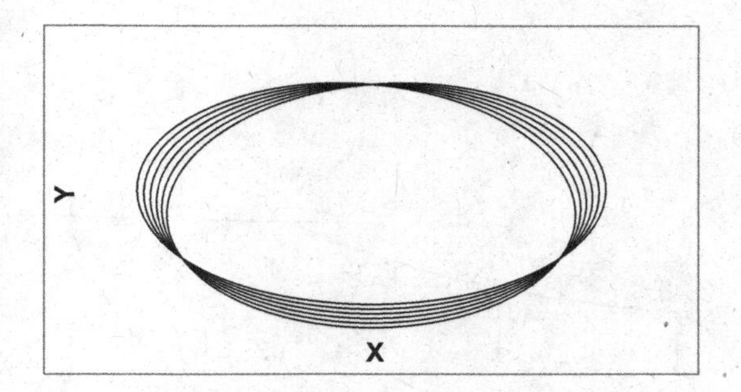

Fig. 14.1. (Pre flattened) NHT = 3 MS as a Concatenation of Trefoil Knots

set of much narrower such strips. Upon twisting and recombining, the leftmost (say) and rightmost subsidiary strips connect to form a torus knot as do the next such pair, and so forth, to form the complete MS in what amounts to a ***concatenation*** of such subsidiary torus knots.

Consider, for example, the MS with ***three*** half twists (Number of Half Twist (NHT) = 3): in this case the knot in question is the ***trefoil*** knot, knotted so as to exhibit three crossovers in a planar illustration. Figure 14.1 shows the *discrete* ***concatenation*** of three trefoil knots (an arbitrary choice; it could have been four or five, or whatever, rather than three such knots) to form such an MS. What look like three nodal points are really just where the ***plane*** of the strip is oriented ***normal*** to the plane of the picture. Finally, concatenation is *one* way to accomplish what is termed the "***framing***" of a knot. Actually, all Moebius Strips (MS), flattened or not and basic or not, can be viewed as "framed" $(2, n)$ torus knots (see below for definition). Thus, the essence of our particle model is seen to reside in such knots and in their associated ***toroidal topology***.

Here's what we mean by a torus knot: In our macroscopic world it's basically just a string (that can be *thought about* as) wound around a torus before its ends are joined. However, a ***caveat***: Our microworld torus here is implicit and ephemeral — it's not really "real". If it were, there would be no way to free it from its knotted imprisonment without drastic surgery! Of course our "string" is not really "real" either in the sense of the tangible thing we're used to in the macroscopic world. However, it can be *thought of* as a

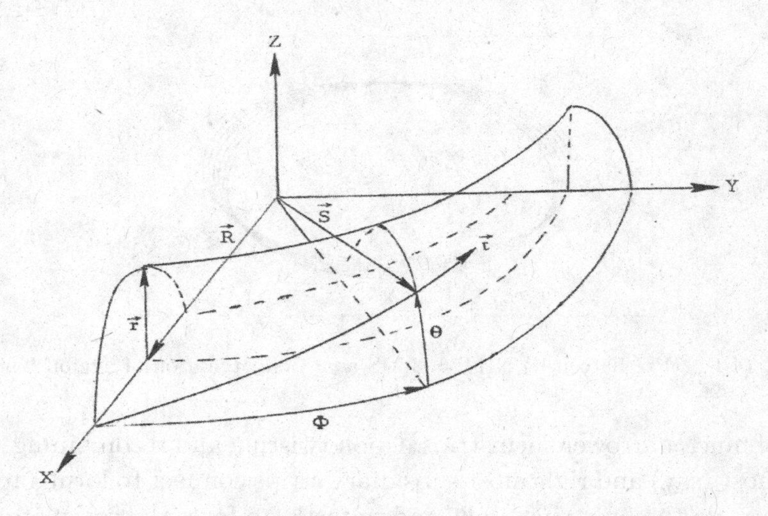

Fig. 14.2. A Segment of a Torus Knot

one-dimensional entity characterized by a ***linear energy density*** — a "***stress***". Nevertheless, it helps to ***assume*** an explicit toroidal geometry in order to perform some analysis, so Fig. 14.2 shows that geometry.

Thus, a point on the toroidal surface is locatable by the vector we see in Fig. 14.2, namely:

$$\vec{S} = \hat{i}x + \hat{j}y + \hat{k}z \tag{14-1}$$

where

$$x = w\cos\phi$$

$$y = w\sin\phi$$

$$z = r\sin\theta$$

$$w = R + r\cos\theta,$$

ϕ is measured in the toroidal core's longitudinal (long way) direction and θ in its meridional (short way) direction.

Restricting the knots to be of genus (2, n) means that, the knot must complete n circuits in the meridional, or θ direction for every

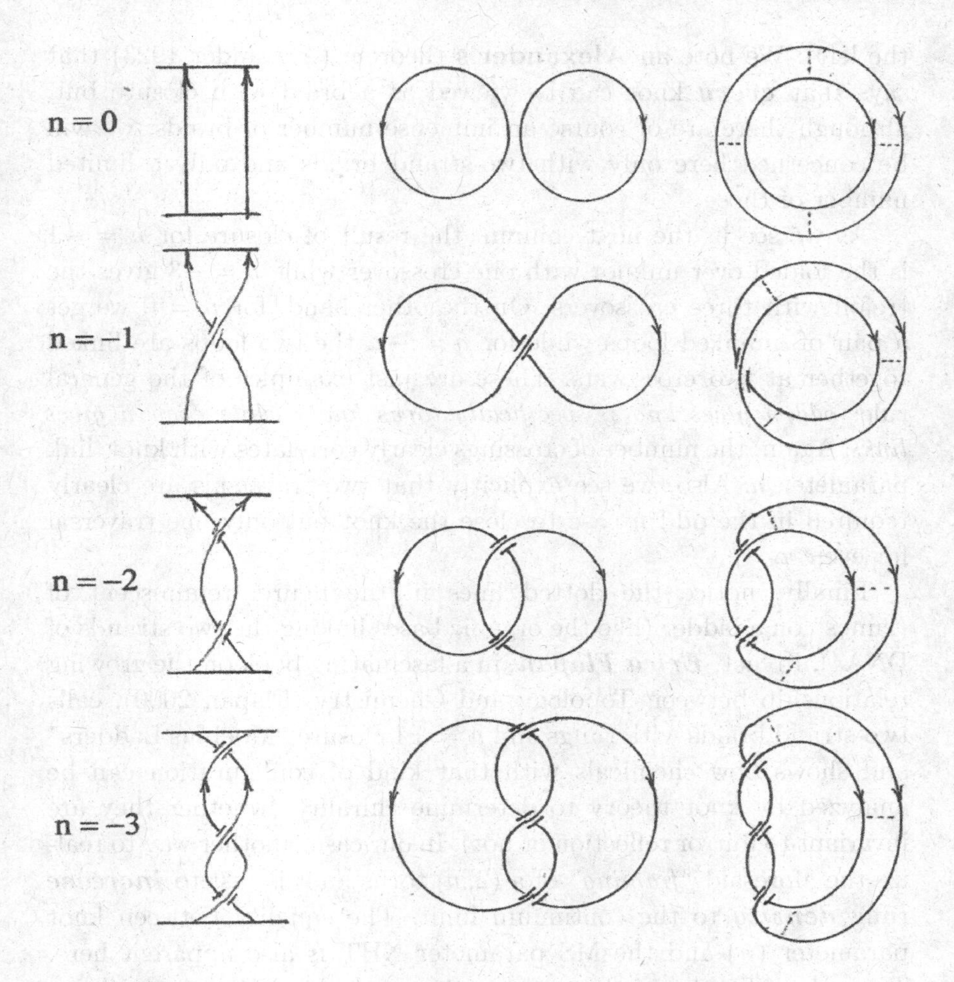

Fig. 14.3. Two-strand Braids with Closure and Partial Framing

2 traversals in longitude, ϕ. We note that it's only for ***odd*** values of n that we get a knot; *even* values create links (which, in this book, translates into ***odd values for Fermions*** and ***even values for Bosons***, respectively). A graphic way to show this is to depict the equivalent situation involving what can be described as two-strand ***braids*** with ***closure***, meaning that the top and bottom of the strands, indexed left to right as depicted in the first column of Fig. 14.3, are connected (the negative sign just means twist to

the left). We note an **Alexander's** theorem (Alexander 1923) that says that ***every*** knot can be viewed as a braid with closure but, although there are of course an immense number of braids, we will be concerned here only with two-strand braids and only a limited number of those.

As we see in the next column, the result of closure for $n = -1$ is the folded over unknot with one crossover while $n = -3$ gives the trefoil with three crossovers. On the other hand, for $n = 0$ we get a pair of unlinked loops while for $n = -2$, the two loops are linked together at two crossovers. These are just examples of the general rule: *odd n gives knots, specifically torus knots while even n gives links.* Again, the number of crossings clearly correlates with knot/link parameter n. Also, we see explicitly that two traversals are clearly required in the odd n case to close the knot but only one traversal for even n.

Finally, notice the dotted lines in the figure, reminiscent of "rungs" on a ladder (like the organic bases linking the two strands of DNA[2]). In fact, ***Erica Flapan***, in a fascinating book on the growing relationship between Topology and Chemistry (Flapan 2000), calls two-strand braids with rungs and $n = \pm 1$ closure "Moebius Ladders" and shows how chemicals with that kind of configuration can be analyzed by knot theory to determine chirality (whether they are invariant to mirror reflection or not). In our case, another way to realize the aforesaid "*framing*" of a $(2, n)$ torus knot is just to ***increase*** rung ***density*** to the continuum limit. The equality between knot parameter (n) and the MS parameter NHT is also apparent here. To quote, "The braided representation makes manifest that all our "particles" belong to the *same genus*, namely the set of framed $(2, n)$ torus knots (or links)". In summary, it would appear legitimate to say that a torus knot (or link) ***is*** a two-strand braid with closure, and a Moebius strip ***is*** a framed two-strand braid with closure!

In any case, in what follows we will be concerned with the nature of Torus knots and their explicit description as Sine-Gordon Solitons,

[2] As per [10], our bosons (even values of NHT) evoke the "cyclic, duplex DNA — two-stranded DNA whose otherwise free ends are connected to each other".

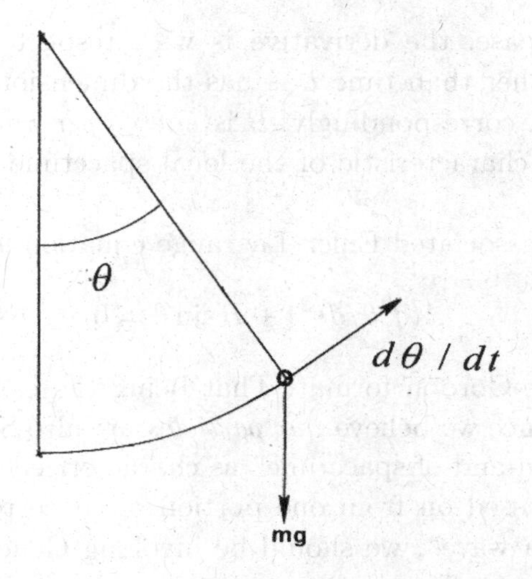

Fig. 14.4. Pendulum (Repeat)

expressed in equation form as

$$d^2\theta/ds^2 + (1/wr)\sin\theta = 0, \tag{14-4}$$

where s, r, w and θ are defined in connection with Fig. 14.2 above. This expression can be derived in more than one way although our purpose here requires beginning with a Lagrangian (which we do in what follows). However, before we do so it is useful to discuss the equation in somewhat elementary terms. We note that it appears in a variety of contexts, but the canonical situation is the time-honored case of an idealized pendulum consisting of a stiff, weightless rod of length l and a weight of mass m constrained to rotate in a plane as portrayed in Fig. 14.4 under the influence of gravity.

As we saw in Chap. 7, the Lagrangian for this exemplary situation is

$$L = (1/2)ml^2(d\theta/dt)^2 + mgl\cos\theta. \tag{14-5}$$

Thus, the message for constructing a Lagrangian for the ***knot*** situation is that we need to start with something like

$$L = (1/2)A(d\theta/ds)^2 + B\cos\theta, \tag{14-6}$$

where, in this case, the derivative is with respect to knot length parameter s rather than time t, A has the dimensions of *energy per unit length* and, correspondingly, B is *energy per unit volume.* Both parameters are characteristic of the local spacetime but, are as yet unspecified.

Finally, the associated Euler–Lagrange equation is then

$$A(d^2\theta/ds^2) + B\sin\theta = 0 \tag{14-7}$$

which is in Sine-Gordon format. That being the case, let us pause and consider: since we believe *our particles* are also Solitons, that is, deformations "in-and-of-spacetime" as characterized in and "continuously being passed on from one portion of space to another after the manner of a wave", we should be invoking General Relativistic relationships. Actually, as far as A is concerned, we have that immediately according to

$$A = c^4/4\pi G \tag{14-8}$$

where c is the velocity of light and G is the gravitational constant, ***nominally*** equal to $6.673 \times 10^{-11}\text{m}^3/\text{kg}\cdot\text{s}^2$.

To specify B we now invoke the notion that contraction of the Einstein equation

$$R_{\mu\nu} - \frac{1}{2}g_{\mu\nu}R = -8\pi G T_{\mu\nu} \tag{14-9}$$

leads to the following relationship between energy and curvature, namely:

$$R = (8\pi G/c^4)T^{\mu}_{\mu} = 2T^{\mu}_{\mu}/A. \tag{14-10}$$

Here $R = g^{\mu\kappa}R_{\mu\kappa}$ is the Curvature Scalar (CvS) and

$$T^{\mu}_{\mu} = T_{\mu v}g^{\mu v}, \tag{14-11}$$

is the Energy Scalar, contracted from $T_{\mu v}$, the ***Energy-Momentum Tensor*** (EMT) as per the preceding chapter.

The Curvature Scalar, is something we can compute on the basis of the definitions associated with Fig. 14.2 and is found to be[3]

$$R = (2/wr)\cos\theta. \tag{14-12}$$

Actually, what we are really interested in here is expressing the Energy Momentum Tensor in terms of the Curvature Scalar, the point of view being that, in this case, ***energy*** is due to curvature (rather than the usual other way around). Combining (14-10) and (14-12) we then have

$$T^{\mu}_{\mu} = (A/wr)\cos\theta,\ \mathrm{kg(m/s)^2/m^3} \tag{14-13}$$

which we set equal to the $B\cos\theta$ in Eq. (14-6) to give us the Lagrangian in the form

$$L = A[(1/2)(d\theta/ds)^2 + (1/wr)\cos\theta]. \tag{14-14}$$

The corresponding Euler–Lagrange equation is thus

$$d^2\theta/ds^2 + (1/wr)\sin\theta = 0. \tag{14-15}$$

At this point we note that although particles are ***visualized*** herein as solitonic disturbances "in and of an otherwise undistorted continuum", there was no mention, to this point, of what it is that does the disturbing. It turns out (and we shall not demonstrate that situation here) that our solitons are very slender and for that condition the actual solitonic disturbance is a very narrow *torsional* ***distortion*** that travels around the putative toroidal core that does so. However, note that there are ***two directions*** of traverse available in the model employed herein and, in effect, only one has actually been invoked, by implication, traverse to the ***right***. Thus, it seems natural to ask whether our *solitonic model* can also accommodate the notion of ***antiparticles*** by invoking traverse to the *left*. Although this was the nominal state of affairs in Chap. 3, it was occasioned by

[3]The derivation is given in the previous book in terms of connection terms but will not be repeated here.

the *requirements* of fusion. Here, we ask whether that requirement is satisfied inherently.

To pursue this possibility, it is expedient to first demonstrate the validity of the well-known solution to the Sine-Gordon equation. First, we note that, in what follows, the dependent variable θ of the Sine-Gordon equation and its first and second derivatives with the independent variable s will be designated by θ_R, θ'_R and θ''_R, respectively with R explicitly signifying traverse to the ***right***). We then begin the demonstration with the solution in the following form:

$$\mathrm{e}^{-\eta s} = \tan\left(\frac{\pi - \theta_\mathrm{R}}{4}\right) \tag{14-16}$$

where $\eta = 1/\sqrt{wr}$.

Differentiating both sides with respect to s then (soon) produces

$$\begin{aligned} -\eta \mathrm{e}^{-\eta s} &= (-\theta'_\mathrm{R}/4)/\cos^2\left(\frac{\pi - \theta_\mathrm{R}}{4}\right) \\ &= 2\eta \sin\left(\frac{\pi - \theta_\mathrm{R}}{2}\right). \end{aligned} \tag{14-17}$$

Solving for θ'_R, (likewise) produces

$$\begin{aligned} \theta'_\mathrm{R} &= 4\eta \sin\left(\frac{\pi - \theta_\mathrm{R}}{4}\right)\cos\left(\frac{\pi - \theta_\mathrm{R}}{4}\right) \\ &= 2\eta \sin\left(\frac{\pi - \theta_\mathrm{R}}{2}\right) \end{aligned} \tag{14-18}$$

and another differentiation then yields the desired result:

$$\begin{aligned} \theta''_\mathrm{R} &= 2\eta \cos\left(\frac{\pi - \theta_\mathrm{R}}{2}\right)(-\theta'_\mathrm{R}/2) \\ &= -\eta^2 \sin(\pi - \theta_\mathrm{R}) \\ &= -\eta^2 \sin\theta_\mathrm{R} \end{aligned}$$

or, as required:

$$\theta''_\mathrm{R} + \eta^2 \sin\theta_\mathrm{R} = 0. \tag{14-19}$$

For the case of traverse to the left, we therefore replace s by $-s$, and dependent variable θ with θ_L. We then begin as in the above with a corresponding replacement for Eq. (14-13), namely

$$\mathrm{e}^{\eta s} = \tan\left(\frac{\pi - \theta_{\mathrm{L}}}{4}\right). \tag{14-20}$$

Differentiating and solving for θ'_{L} produces:

$$\theta'_{\mathrm{L}} = -2\eta \sin\left(\frac{\pi - \theta_{\mathrm{L}}}{2}\right) \tag{14-21}$$

instead of Eq. (14-15) whereupon, with a second differentiation and corresponding simplification we get

$$\theta''_{\mathrm{L}} - \eta^2 \sin\theta_{\mathrm{L}} = 0 \tag{14-22}$$

which is Eq. (14-15) with a minus sign instead of a plus sign between the two terms or, put another way,

$$-\theta''_{\mathrm{L}} + \eta^2 \sin\theta_{\mathrm{L}} = 0. \tag{14-23}$$

Note that the first term is traceable back to kinetic energy and the second term to potential energy. However there is now a ***minus sign*** between the two terms rather than a ***plus sign*** as for the Sine Gordon original solution for traverse to the right. So: If we actually do the tracing we see that the ***Lagrangians*** for the two situations must be ***different***; for traverse to the right it is K−V but for the traverse to the left it is K + V!

We can substantiate Eq. (14-20) by showing how a leftward traverse implies a negative mass. We begin by multiplying Eq. (14-24) by $Ar(v/c)^2(\Delta\ell)$ where $v = ds/dt$ is velocity along the knot. The result, again to first order in r/R, is

$$[(A/c^2)\sqrt{\mu Rr}](r\ddot{\theta}_R) + [(v/c)^2\sqrt{r/R}](A/2)\sin\theta_R = 0, \tag{14-24}$$

which is in the "Newtonian dynamic" form

$$m_{\mathrm{R}}a_{\mathrm{R}} + F_{\mathrm{R}} = 0 \tag{14-25}$$

where

$$\begin{aligned} m_{\mathrm{R}} &= (A/c^2)\sqrt{Rr} = c^2 g/4\pi G(\mathrm{kg}) \\ a_{\mathrm{R}} &= r\ddot{\theta}_{\mathrm{R}}\ (\mathrm{meters/sec}^2) \\ F_{\mathrm{R}} &= (v/c)^2\sqrt{r/R}(A/2)\sin\theta_{\mathrm{R}}\ (\mathrm{joules/meter}). \end{aligned} \tag{14-26}$$

Note that acceleration varies as $a_{\rm R} = \left(\frac{v^2}{\rm R}\right) \sin\theta_{\rm R}$ and that mass is proportional to the ratio of The Metric (as defined above) to the Gravitational Constant G (which is essentially the inverse of the linear energy density term A — see Chap. 6, this Section).

On the other hand, multiplying by that same expression, $Ar(v/c)^2(\Delta\ell)$, we get an equation analogous to the "dynamic equation"

$$-m_{\rm R}a_{\rm L} + F_{\rm L} = 0 \tag{14-27}$$

where, as per the above (but with modifications):

$$m_{\rm R} = (A/c^2)\sqrt{Rr}\ \text{(kg)}$$
$$a_{\rm L} = r\ddot{\theta}_{\rm L}\ \text{(meters/sec}^2\text{)}$$
$$F_{\rm L} = (v/c)^2\sqrt{r/R}(A/2)\sin\theta_{\rm L}\ \text{(joules/meter)}.$$

That is, $m_{\rm R}$ is the same as the m in Eq. (14-26) but $a_{\rm L}$ and $F_{\rm L}$ both now incorporate $\theta_{\rm L}$ rather than θ. Thus, in order to recreate the ***format*** of the dynamic equation, that is

$$m_{\rm L}a_{\rm L} + F_{\rm L} = 0, \tag{14-28}$$

we need to ***define*** the mass term for ***leftward*** traverse as $m_{\rm L} = -m_{\rm R}$.

In summary we see that the ***sum*** as well as the ***difference*** between the two energy terms is a valid concept at least as applied in the case treated above. The two formulations may be viewed as being ***complementary*** manifestations of an ***enhanced Action principle***. Or, conversely, the Action principle may be viewed as a particular instance of allowable manifestations of the Principle of Complementarity of which, we recall, Noether's theorem is also a manifestation!

15

An Authoritative Perspective

In April 1950, Albert Einstein wrote "On the Generalized Theory of Gravitation" (Scientific American 1990). Here's a snippet of what he had to say about electrodynamics and Special Relativity: speaking of Michael Faraday "unencumbered by the traditional way of thinking, he felt that the introduction of the 'field' as an independent element of reality helped him to coordinate the experimental facts".

Whereupon Maxwell "made the fundamental discovery that the laws of electrodynamics found their natural expression in the differential equations for the electromagnetic and magnetic fields. These equations imply the existence of waves whose properties correspond to those of light". And, finally: "Maxwell's equations for empty space remain unchanged if the spatial coordinates and the time are subject to Lorentz Transformations".

Unwittingly, Einstein also had something to say about how General Relativity impacts our Alternative Model of the Elementary Particles, to wit:

"According to general relativity, the concept of space detached from any physical content does not exist. The physical reality of space is represented by a field whose components are continuous functions of four independent variables — the coordinates of space and time.

Since the theory of general relativity implies the representation of physical reality by a ***continuous*** field, the concept of particles or material points cannot play a fundamental part, nor can the concept of motion. The particle can only appear as a ***limited region in space*** in which the field strength or the energy density are particularly high."

16

And an Inference Therefrom

It seems to me that the Elementary Particles that appear in Chap. 9 above are just exactly what the good Professor is talking about! Perhaps you recall the phrase I quoted from one of my earlier publications describing my elementary particles as solitons existing "in-and-of the fabric of spacetime". Precisely! And each such "particle" moves through space as a wave which is what a ***soliton*** does — continually forming and reforming of that fabric of spacetime! Not the usual concept of particulate motion. Furthermore, the set of elementary particles epitomize the Principle of ***complementarity***, so we don't have to worry about their legitimacy!

So there you have it; in a way, a sort of "Ultimate Reality". The only "real" entity is the "something rather than nothing" of ***spacetime itself*** and everything beyond that exists as a localized distortion in one way or another of it and the organization and interactions of such.

Of course that's not all there's to it: at a minimum our elementary particles have to be capable of giving rise to all the complexity of the real universe meaning they must at least provide for a ***taxonomy*** of higher organization and for interactions between the residents thereof. We saw a hint of that in Chap. 3 and we shall expand upon it a bit later. Moreover, there remain some very basic questions that just the existence of a set of elementary particles can't yet answer.

We need some more background, in fact very important background. So let's get going on that.

17

An "Indigenous Parallel" to the Higgs

This chapter is a condensed version of the notion discussed in the previous book (Avrin 2015) that each basic AM fermion experiences "a situation reminiscent of the symmetry-breaking topology that preconditions the ***Higgs*** mechanism. The complete Higgs mechanism is, necessarily, rather complicated; without it the Standard Model (SM) would not be able to attribute mass to the elementary particles of the model without violating some of its basic theoretical underpinnings, primarily SU(2) gauge symmetry and the efficacy of renormalization. However, it is not necessary to invoke the entire mechanism here because, in the first place, SU(2) symmetry in the AM is ***inherent*** to particle structure and, furthermore, renormalization is not an issue. What is ***basic*** to the SM's Higgs mechanism, however, is the need to introduce a symmetry-breaking version of the ***potential energy*** that each particle encounters. That potential is also of interest here because of the way in which symmetry breaking was introduced into the AM, that is, by way of the initial implicit ***assumption*** of toroidal topology and the ***reduction***, as per the above, of the value of linear energy density, A, of spacetime in the neighborhood of the MS.

For ***comparison*** purposes, the previous book adopted a version of the Standard Model's Higgs potential that corresponds to the commonly invoked ***Landau-Ginsburg*** model, the one based on the spontaneously broken symmetry encountered in ferromagnetism, namely

$$V = \mu^2|\phi|^2 + \lambda|\phi|^4, \tag{17-1}$$

which, without going into detail, may be described as a "Hill and Valley" characteristic, that is, a central "bulge" surrounded by a depressed region.

In comparison, what was described in as an "indigenous version of the Higgs-like potential" is arrived at in a radical different way. Before we display it, we can gain some important insight if we consider the following very short but *not at all inconsequential* ***Theorem*** (Referred to as the NI Theorem):

Suppose we have two variables, x and y which interrelate according to each of two equally valid relationships;

$$1.\ y = ax^n$$
$$2.\ y = bx^{-n}.$$

Summing the two, multiplying by x^n and "shuffling" we obtain the equally valid relationship,

$$V \equiv -b = -2yx^n + ax^{2n}. \qquad (17\text{-}2)$$

Seeking extrema we find another "Hill and Valley" characteristic with a maximum of $V = 0$ at $x = 0$ for ***relationship #2*** and a minimum of $x = (y/a)^{1/n}$ corresponding to ***relationship #1***. In short, we also have a ***Hill and valley*** characteristic for the NI theorem, whereupon making the proper associations shows the SM Higgs characteristic and that of the NI theorem to be ***identical***.

The theorem also allows us to construct that "indigenous parallel" to the Higgs but it takes a little more work. To begin with we note that, with $n = 2$, the ***first*** relationship in the theorem obtains if we set $y = R, x = m$ and,

$$a = 9\pi^2 G^2/c^4 N^3 L_{PL}. \qquad (17\text{-}3)$$

Where

R is the toroidal radius
G is the gravitational "constant" (not specified here)
C is the speed of light
L is the Planck length.

For the ***second*** relationship we have adopted an expanded approach to the mass vs. size issue, one involving considerations of ***quantum*** mechanical behavior, albeit of the "old quantum theory", in the expectation of a fuller quantum mechanical treatment at a later date. We refer to what in was called "Einstein's ***Ansatz***", the one that maps the quantized trajectory of a particle moving in a central field of force onto a torus in order to resolve momentum ambiguities. Actually, the plan here is to apply the Ansatz in ***reverse***; that is we ***postulate*** the existence of a central field of force that acts upon the solitonic distortion we described above, the implication for our case being that there exists a putative centrally located ***source*** for such a field and, correspondingly, discrete orbits that might be identifiable with the toroidal loci of our model's basic fermions.

In other words, we shall now consider the simultaneous ***validity***, as in the above, of ***two*** completely independent models, one the ***solitonic*** model discussed in the foregoing and the other an ***orbital*** quantum model, hereinafter to be known as the "Converse Einstein/Bohr (***CE/B***)" model. As we know, a hundred years ago, Niels Bohr formulated a model of the Hydrogen atom that predicts that element's discrete line spectra (Bohr 1913). Bohr considered the electromagnetic attraction

$$F_E = e^2/R_o^2 \tag{17-4}$$

between a proton in the nucleus and an electron in orbit as balanced by the centrifugal force experienced by the electron due to its orbital velocity. Bohr also postulated that angular momentum is quantized in multiples of Planck's constant according to

$$mvr = n_o h/2\pi. \tag{17-5}$$

The combination of these relationships results in an expression for the ***radius*** *of the orbit*, namely

$$R_o = h^2 n_o^2/4\pi^2 me^2. \tag{17-6}$$

For the case of our (CE/B) model we should expect a ***similar*** relationship, except that the force of attraction would not be electromagnetic and, of course, we are concerned with much smaller scales. As shown in, solitonic distortion in our particle model is concentrated in a small, mobile region of longitude to which we now attribute the *totality* of particle mass. Although we have not yet discussed its kinematics, assuming that the Bohr analysis is applicable implies that the implicit ***velocity*** of circulation of such a "pseudo particle" is ***immaterial***. However, if the force of attraction between the putative central "entity" (we are not yet prepared to call it a particle) and our solitonic pseudo particle were ***gravitational*** rather than electromagnetic, we should replace the force, F_E, by the ***gravitational*** force

$$F_G = GMm/R^2, \quad (17\text{-}7)$$

where M is the mass of a putative centrally located entity with the result that the relationship for "***orbital***" radius R becomes

$$R = h^2n^2/4\pi^2 GMm^2 \quad (17\text{-}8)$$

and amounts to just replacing me^2 by GMm^2.

We now assume the ***simultaneous*** validity of our two models, the solitonic behavior as per the above repeated here for reference,

$$m = (c^2/3\pi G)N^{3/2}\sqrt{RL_{PL}}$$

and that for the CE/B model, namely Eq. (17-8), or, more compactly,

$$1.\ R = \alpha m^2$$

$$2.\ R = \beta/m^2 \quad (17\text{-}9)$$

where, with $n = 1$

$$\alpha = 9\pi^2 G^2/c^4 N^3 L_{PL}$$

$$\beta = h^2/4\pi^2 GM = h^2 A/\pi c^4 M$$

which is also ***in the form*** of our little NI theorem! In other words, we now know how to realize what we might now refer to as the Alternative Model's ***version*** of a ***Higgs*** potential. There are various ways to express it. One way, in analogy with the Landau-Ginsburg model that we wrote down for comparison purposes, is as the potential

$$V_{\text{comp}} = -\eta^2|\phi|^2 + \varsigma|\phi|^4, \tag{17-10}$$

where, to make things a little more *physical* we have used

$$\begin{aligned} A &= mc^2/2g \\ \eta^2 &= 4Rg^2/c^4 \\ \varsigma &= 4\alpha(g^2/c^4)^2 \\ g &= \sqrt{Rr}. \end{aligned}$$

To recapitulate, the main thesis of our Alternative Model (AM) of the elementary particles has been that it constitutes an actual, *physical* ***manifestation*** of the taxonomy, interactions and attributes of the Standard Model (SM) but without recourse to the latter's quarks, gluons or color. At some point in its development, the SM was considered by its participants to be a complete theory with one notable exception; it was impossible to endow its particles with mass and retain gauge invariance and renormalization efficacy unless the symmetry of the potential energy they experience is broken. Ergo, the notion of a potential energy field — the so-called Higgs field — such that the potential energy a particle experiences depends on its location as measured in the ***coordinates of that field***.

We emphasize that such coordinates are ***not physical*** — that is they do not have the dimensions of ***space***. As noted at the beginning of this section, the formal expression for the field is sometimes described as being modeled after the symmetry-breaking field of ferromagnetic magnetization theory. The field may be characterized by its "hill and valley" ***topography*** — that is, a centrally located mound surrounded by a radially-symmetrical trough. As we have seen, it consists of a quadratic term that describes the descent down from the hill and a quartic term that reverses the downward slope and climbs up out of the valley. Symmetry breaking occurs because location at the top of the hill is only conditionally stable and the particle's preferred location in field coordinates is ***somewhere*** at or near the bottom of the valley but in ***any radial*** *direction*.

The identical field is assumed to *exist at* ***any location*** *throughout space* so that a particle moving through it will be affected in a manner sometimes rhetorically described as a kind of ***drag*** on its motion. It seems fair, at this point, to say that the Higgs field of the SM may be viewed primarily as an ***attribute of space*** rather than of any individual particle, even though not all particles are affected by it to the same extent. Finally, the culmination of a rigorous search, using the particle accelerator at CERN — the European Organization for Nuclear Research, to collide protons together at high energy. has apparently identified a hitherto unknown particle which has the attributes associated with the quantum of the Higgs field, the so-called Higgs boson, thus, it is maintained, removing the last obstacle to the complete validation of the SM.

However, what has emerged in the foregoing, and in particular in this section, is that a *real, physical* ***parallel*** to the hill-and-valley topography of the Higgs field can be associated with each elementary AM ***particle*** by permitting the gravitational field in its ***interior*** to vary in a particular way in ***particle coordinates***. The Higgs-like characteristic was arrived at herein by assuming the simultaneous validity of two independent relationships between particle mass and size: one associated with our Sine-Gordon ***solitonic*** MS and the other Niels **Bohr's** model of the *Hydrogen atom* as adapted to a ***gravitational*** rather than an electromagnetic force. Recall, if you will, our repeated insistence that what may be described as the "last" for our Sine-Gordon, ***Moebius strip*** elementary particles, namely an actual ***torus***, *does* ***not*** *really exist.* What we see instead is the friendly ***valley*** of this gravitational characteristic providing the necessary ***toroidal*** *topology* for our MS to ***occupy*** while the ***hill*** provides a putative *central* ***field*** *of force*, not initially considered as part of the model, for the MS to take shape as per Einstein's Ansatz.

And finally, one more thing: just as our matrix m may be regarded as the generic representative for ***quaternary complementarity***, so can our little NI theorem be regarded as the generic representative of ***binary*** complementarity.

18

Undulation (or Undulatority)

About the title: according to the verb "undulate" is defined as "to move with a wavelike motion". Also, as used in this book, "Complementarity" is basically a discussion of the verb to "complement". So, can "Undulatority", which is hereby defined to mean discussions of wavelike motion, pair with "undulate" in the same way that "complementarity" pairs with "complement"? Why not?

Anyway, this chapter is, indeed, about waves. Waves are fascinating entities; all kinds of waves — in all kinds of media[1]; they are ubiquitous and beyond just useful, even necessary. Mankind has been aware of wave phenomena for countless millennia. Throw a stone (not a great big rock!) into a pond and a familiar series of concentric ripples is excited to spread out along the surface. Thus, the energy that disturbance introduces into the water at one location is spread out over the pond.

There is of course another meaning associated with the Infinitive "wave"; people do it! That is, we wave by implementing some kind of undulatory motion. It happens all the time, mainly as a form of greeting or the acknowledgment of parting, right? (Although my wave is often more like an incomplete salute). Even infants are encouraged by their parents to wave "Bye, Bye" with a more-or-less standard up-and-down motion of the hand with the palm in the down position.

[1]Or none: You may recall where, in the Preface, the Schrödinger approach to Quantum Mechanics was characterized as emphasizing its "undulatory aspects". And though Schrödinger's equation resides for analytical purposes in Hilbert space, I don't believe its wavelike solutions live there as well.

Once the tykes learn it, it may become ritualistic and hard to squelch!

And speaking of ritual, her Majesty, Queen Elizabeth of the house of Windsor, has become expert at majestically acknowledging the plaudits of her subjects during royal outings. The royal wave in this case is a fascinatingly-demure left-to-right slice of the air by the royal palm which faces her Majesty in a vertical plane parallel to the path of the Royal Rolls — it's a kind of inverted Pendulum with, in this case, the forearm rotating with the palm while the elbow is stationary and in the down, fulcrum position. I trust there's no need for further description. Perhaps you can verify the phenomenon as it occurs on the evening news broadcast.

Alright, that's enough of "ritual"; we hadn't really finished with "ripples". How should we describe them so as to fit in with the rest of this chapter? Here's one way: a "ripple" is basically just a localized ***curvature*** pretty much confined to or close to the surface of a body of water, and moving outward from an initial disturbance, for example, that pond with parameters dominated by surface tension. That ought to be good enough for now but the question is really what do we mean by "ripple" in general, in fact what constitutes a "pond" in general. Well, for example, it turns out we can increase the overall scale of the "pond" a lot. How much — to a lake? A bay? How about an Ocean? I must be joking, right? It would take a rather large rock in that case! Actually, as the astronomers and geologists, etc. told us, something like that really did occur: About 65 million years ago an immense asteroid slammed into our planet in the Atlantic neighborhood east of what's now Central America, thus initiating the planetary conditions that wiped out the Dinosaurs. And made it possible for us — eventually — to be here talking about it.

But that's not what I have in mind. What I want to talk about at this point, believe it or not, are ***Tsunamis***, something we hear a lot about when one of them strikes the shore at various points around the Pacific Ocean, especially one of the Islands of Japan. In fact, as is well-known and not surprising, "Tsunami" is a Japanese word. What is not well-known is that, if we stand back and think about it, a Tsunami may be viewed as kind of a set of "ripples" in the form of very large waves that also distribute the energy imparted to the ocean by an external disturbance. Of course, the whole phenomenon is clearly scaled up by many orders of magnitude and the source of the energy is not a big rock splashing down from above — instead it's an ***earthquake***, one that occurs well below the ocean floor. And the waves that result are, of course, not tiny ripples on the surface either, but large, sometimes huge, waves whose repetition rates are usually discussed in terms of ***minutes*** between successive waveforms rather than as a frequency of occurrence. Nevertheless, the comparison is legitimate; we do have a source of ***energy*** and its ***dissipation*** is indeed enabled by a set of ***waves***. All of which is naturally of small comfort to the residents of Japan living on or near where the Tsunami waves strike.

In any event, that's about as large a wave structure as we can find on the surface of the planet; for a phenomenon of even larger scale we have to go on up into the ***atmosphere***. Perhaps you've read the article by **James Fleming** (2017) in a recent issue of ***Physics Today*** about ***Carl-Gustaf Rossby*** who among, other things, is known for the initial study of ***atmospheric planetary*** waves with "wavelengths on the order of 5000 km with, typically, four or five ridges and troughs at any one time encircling the North Pole in sequence at or above the 40th parallel". These are ***volumetric*** waves (that is, they exist in 3D) in the atmosphere and no distinct "surface" is involved.

However, in the figure that accompanies the article, the "ridges and troughs" translate to northern and southern ***excursions*** relative to the Pole — a kind of "***lobe structure***" known, not surprisingly, as "***Rossby Waves***" which appear to be more-or-less bounded on the south by what is referred to on the evening television weather report as a ***jet stream*** that approaches and retreats from the pole even as it encircles it. The overall system covers enough chilly real estate up there so that the illustration in the article clearly shows the influence of the well-known "Coriolis effect" known to be associated with the rotation of a spherical planet like ours. In any event, as we know, the overall system certainly transmits a lot of energy, ultimately traceable to planetary motion and the particular weather effects I know little about, beyond what I absorb from the evening news.

But getting back to waves on the sea — even excluding Tsunamis, such waves can of course get bigger than ripples on a pond — a lot bigger, in terms of both amplitude and wavelength and when that happens, surface tension is no longer much of a factor; it's the ***wind*** that generally stirs up the surface. I live in a town on the hilly Palos Verdes peninsula about 25 miles south of Los Angeles, California at an altitude of 1000 feet above sea level. It's about a mile to the western shores as the crows around here can fly if so minded, but about three miles down to the south and the road encircling the peninsula. I used to run down there on Sunday mornings. And back up of course but those days are gone forever. Nowadays my wife and I like to drive down to a park branching off the road to a cliffside lookout where we can see the waves rolling in to crash on the rocks below. These are certainly not ripples; they are often the remnants of the majestic, deep ocean waves that can develop far away, out on the ocean. If the waves are under the influence of a strong, prevalent wind we can see the approach of quite large amplitude waves with long wavelengths.

Many decades ago I got a chance to encounter that kind of wave. Just after New Years of 1945, I was treated to a trip on the Atlantic from New York to Le Havre. If I recall correctly, the name of the ship was the General Sherman. It was not a very big ship for which it compensated by not providing very luxurious accommodations for passengers either! Although the trip was free, on the first day, as we steamed slowly out through the harbor, I was presented by the organizers of the trip with a long fork with which I was instructed to serve hot, steaming bacon for the breakfast of a couple hundred hungry young men (like myself). However, upon discovering, as breakfast wore on, that the fork was not long enough to allow me to maintain a healthy attitude so close to that steaming bacon I decided right then and there to discontinue the service; I dropped the fork and dashed upstairs to take some deep breaths and ended up spending the rest of the morning on deck where the healthy attitude soon returned.

Apparently, I was not missed and being up in the air was such an enjoyable experience that I spent much of each day thereafter up on deck. The weather was generally blustery, but I actually enjoyed the wind and the motion of the ship, noting that the numbers painted on the bow often went up and down some twenty or thirty feet or so over the course of five to ten seconds as I recall. Had I known the ship's speed (and had I been so inclined at that stage in life) I might have been enabled to estimate some wave parameters especially the wavelength of the magnificent disturbance the ship was traversing. In fact, knowing the distance from the USA to France and the elapsed time thereto I might have been able to estimate the speed had there not been a complication; it turns out that the ship was not taking a direct course to Europe; in fact it was traversing a rather zig-zag pattern which one might have assumed was calculated to confuse chance observers. Which of course it was!

Well, to make a long story a bit shorter, the return trip to New York (over a year later) was direct. It took ten days and upon measuring the distance traversed (on my globe of the world) I have estimated the wavelength of the waves we encountered previously as maybe a couple hundred feet which is not really extreme; we know that considerably larger wavelengths occur; I just thought you might like to know how it feels going through big sea waves while standing on the deck of a not very big ship in the winter making believe you're a seaman!

By the way, my understanding is that wave motion on the sea actually occurs in an essentially vertical plane that contains the apparent travel motion of the wave, in fact, basically a quasi-circular motion of the water molecules involved, as seen in that plane. In contrast, sound waves ***in air*** are strictly ***longitudinal*** — the air molecules are induced by the emission of sound to move back and forth along the direction of travel rather than crosswise or up and down, which is fortunate because the first physiological entity we possess that utilizes the auditory information contained therein — our ***ear drums*** — seem to be designed to respond to that kind of excitation! Of course music is also composed of that kind of "information" and it may or may not have occurred to you that the reason the ear "drum" is so named, I suspect, is because of its resemblance to a ***real drum*** — a mechanism generally involving a taut ***membrane***!

And did you know that ***sound*** can also proceed through water in the same way? Of course you do; creatures of the sea have been using it to navigate, communicate, to hunt or evade interception a lot longer than we have existed as a species. But mankind's ingenuity has come up with a technologically similar device. It's called **Sonar** (a takeoff on Radar) and the sound is generated and received by what are known as ***Transducers*** wherein, in transmission, a mechanical motion, generated electrically, is converted into a longitudinal waveform by a ***membrane*** of sorts and upon reception the process is reversed. Many decades ago, I had a summer job where I employed instrumentation to barge on a lake to measure Sonar Transducer beamwidths and bandwidths just like in radar. A rather easygoing job but useful I was told.

By the way, the eardrum may of course also be viewed as a transducer, in this case one that, I understand, introduces sound energy excitations to a complex set of hair-like mechanisms that are somehow responsible for generating the frequency structure of the electrochemical signals the explicit auditory sections of the brain translate as auditory information. That's as far as I'm able to explain the subject. Personally speaking, although my wife has frequency-tuned hearing aids, we are constantly asking each other to repeat what we said!

Wavelike phenomena can of course be realized in all kinds of other media by the stimulation of the underlying molecular or atomic structure therein to mechanical motion. The kind of membrane we're concerned with above might be, say, on a kettle drum in an orchestra, in which case a thin, essentially two-dimensional sheet (composed in days gone by of animal hide) tightly stretched so as to supply vibratory energy as in the above and the excitation is supplied by an external mechanical force (in the orchestra, a drummer wielding that padded drumstick). Or, in the case of Sonar, the force would be acting against the electronically induced resistance of the otherwise quiescent molecular/atomic arrangement of the transducer membrane being disturbed.

As we know, the simplest, most straightforward ***wave equation*** associated with some vibratory or undulatory system just pits the second derivative of an extensive variable, say x, against x itself. The resulting solution of the equation is then just a ***sine wave*** whose second derivative is its negative as per the statement of that wave equation. On the other hand, the two-dimensional array of molecules/atoms making up a membrane requires more complexity in a wave equation and, in the resulting two-dimensional patterns of solutions, I seem to recall the presence of at least Bessel functions among other things. That being the case, for the purposes of this book, I see no need to get into that complexity. But one thing we should note about the actual physics involved before leaving this subject is that the resistance of adjacent atoms to separation is indeed ***electronic*** in nature which suggests that we ought to segue at this point into the subject of ***electromagnetics***.

However, if you don't mind, I would first like to discuss one other topic, namely ***Solitonic*** phenomena in terms of their realization in **water**. Although there are a plethora of types of Solitons and the media in which they can occur, the elementary particles that we described as Sine-Gordon distortions of spacetime being one very special example, much more common are the simple motions of large scale configurations in relatively large bodies of water. Many years ago, when we were still doing that sort of thing, my wife and I went on a leisurely sightseeing motor trip of Eastern Canada. We landed in Halifax and proceeded up the Atlantic side of Nov Scotia and Cape Breton Island, having a wonderful time.

On the way back we mostly had the Bay of Fundy (BoF) on our right. As you may know, the BoF is noted for gigantic tides. Well, when we stopped one day for lodging and dinner it was at a town on a river and we were told that we just had to see the tidal "Bore" on the river that evening. In retrospect, the town might have been Truro in which case the river would be the Salmon but, in any case, sure enough, there it came, leisurely but unhesitatingly sweeping up from the bay, a distinct, wedge-like conformation, in this case only about a foot or so high but extending all the way from bank-to-bank across the river and proceeding without pause up river from left to right as we stood on the south bank. We note that the tide initiates the bore but after it, itself, diminishes and the bore is sufficiently developed, the tide disappears and the bore proceeds on its own! In other words the tide may be regarded as ***launching*** the bore.

What we saw was, I believe, another example of the classic ***soliton*** first viewed in the early, mid-19th century by the British Engineer ***Scott-Russell*** (usually recognized as the discoverer of Solitonic phenomena) as he rode along the banks of a canal in which a horse-drawn canal boat had somehow ceased moving, thus ***launching***, as a Soliton, the mound of water it had been pushing along as some kind of bow-wave (Eilbeck 1995). Although the overall phenomenon is quite complex, involving canal width and depth as well as the boat's configuration and speed, in essence what Scott-Russell saw proceeding up the canal was pretty much what we saw on the river. There were of course no horses that we saw on our riverbank but what was supplying the push in our case was of course the bay's inexorable evening tide. In fact, we subsequently viewed basically the same kind of "bore" from a bridge over the river that flowed through the city (it might have been Moncton) we were staying in one night on the New Brunswick side of the bay a couple days later.

Well that's about as much as we're going to talk about in terms of the waves to be encountered here in the familiar, real space of atoms and molecules, that is to say in terms of undulations in air or water or in solid materials, etc. One thing we should mention is that, in each of the situations we described above, the Soliton was "hemmed in" you might say, by the banks of the river or the canal. In contrast, as stressed repeatedly, the ***solitonic nature*** of the elementary particles we introduced in our Book of Reference exist 'in and of' Spacetime itself, simply by virtue of their self-sustaining, geometric nature.

So now, how about ***electromagnetic*** (**EM**) ***waves***? Do they exist in "real" space and if not, in what do they exist? Of course, Maxwell's equations predict their existence mathematically (see Chap. 9) but do not specify the medium for their propagation. And, we recall, a lot of people, in mainly the 19th century, postulated that it had to be a mysterious substance called the ether. Of course ***Albert Einstein*** put a stop to that but actually, although EM does ***occur in*** space (or, after ***Minkowski***, in ***Spacetime***), it's not ***usually*** discussed as a ***disturbance of*** space (But, see below). There are in fact two ways to talk about Electromagnetic radiation (actually three ways if we count Geometric Optics!) the ***classical*** way and the ***Quantum*** way.

In the classical way, that is in accordance with Maxwell's equations, we view an Electromagnetic wave as consisting of mutually transverse electric and magnetic ***vectors*** growing and subsiding in consonance in planes normal to the direction of propagation. Previously we talked a bit about the way radar signals are transmitted and received by antennas. Actually, what we talked about was how the radar ***beam*** is formed to transmit or receive an electromagnetic waveform. A straightforward way to do that is to use a "dish-like" reflector, which upon transmission, is illuminated by what amounts to an "***electromagnetic transducer***", induced to convert the back and forth ***motion***, not of atoms nor of molecules, but of ***electrons***, into electromagnetic radiation. And, conversely, upon reception, of such radiation back into an electronic current. That transducer is generally viewed as a ***dipole*** radiator that can consist of just a short, slender, rodlike metallic conductor excited at one end by the electronic signal. And, since electrons are associated with a ***spin*** of $\hbar/2$, electromagnetic radiation should be known as a ***spin 1/2, dipole radiation***, right?

But it's not! It's known as ***Spin 1*** dipole radiation. So why is that? Well, it's a fairly long story but we can compress it herewith. First of all, Spin is regarded as a Quantum Mechanical entity. ***Max Planck***, you may recall, was concerned with the thermodynamic stimulation of ***surfaces*** inside a "Black body" to emit electrodynamic radiation (APS 2002). The surface in this case also functions as a transducer that, prior to Einstein and Bohr — which is to say, in the ***classical picture*** — converts heat energy into the motion of electrons, a motion that results in the radiation of electromagnetic waves. Of course after ***Planck's*** epiphany and ***Einstein*** and ***Bohr's*** insights, to say nothing of the contributions of ***de Broglie*** and ***Compton*** (Famous Scientists 2016) and the other heroes of the quantum revolution (Kleppner and Jackiw 2000), we can now talk about the radiation of the ***discrete quanta*** of electromagnetic radiation known as ***Photons*** with ***spin 1 and mass 0***.

So, where do those numbers come from? Well, let me bring to your attention once again our ***Alternative Model*** of the Elementary Particles in which a ***photon*** is modeled as the ***fusion*** of an ***electron*** and a ***positron***. In this model, ***spin*** is additive which, in this case produces a spin of 1, and where ***mass (and charge)*** are each added ***algebraically***. Also, as you may again recall, since the mass and charge of the positron are the ***negatives***, respectively, of those of the electron, the mass and charge of the photon each disappear and our photon proceeds through spacetime as a ***massless, spin 1*** particle with ***no charge*** at all!

Which, you may or may not be somewhat surprised to realize, brings us, finally, to ***Gravitational radiation,*** which, it turns out, is characterized as a ***spin 2, Quadrupole*** phenomenon. That sounds like it may take twice as much discussion as in the dipole case to really clarify that "Characterization"! Or maybe not; let's see. In the first place, gravitational phenomena are intimately associated with the ***curvature*** of ***spacetime***. And, of course, the source of all gravitational discourse is generally acknowledged, to be Einstein's magical equation in which local ***curvature*** is equated to local ***Stress-Energy***.

Here's a much simpler example of that kind of thing: Archery! You just pull on the string and the bow develops a curve. The energy you expend in pulling is now stored as ***stress-energy*** in the bow. Then, when you let go the string, the bow relaxes precipitously and there goes the arrow, full of commensurate ***kinetic*** energy! So, now harking back to our friendly example of ripples in a pond, we see that ***any*** general kind of modification in the local stress-energy of space should be accompanied by a corresponding ***quadrupole curvature*** to be balanced out by the emission, in this of case, of some kind of ***quadrupole wave*** structure — generalized "***ripples***" so to speak. In summary, if one were to ask the rhetorical question: what is it that's being stressed? The answer, of course, is it's not a bow and it's not the surface of pond water; it's **space itself** and we find it responding with ***local curvature***! And what this gravitational ***radiation*** really amounts to is the ***propagation*** of a wave of curvature as impressed upon that otherwise ***quiescent spacetime***!

So, the problem is then how to use that magical Einstein equation so as to formulate some kind of quadrupole wave equation. Many have tried and the general consensus is that the best one can do is to simplify to the point that nonlinearity is removed or at least doesn't complicate the analysis too much. In the end, there are a number of things everyone seems to agree with[2], but what we shall ***focus*** on here is the ***effect*** a gravitational wave can have as it passes through and temporarily excites that "otherwise ***quiescent region*** of spacetime".

Our discussion revolves around a couple of figures shown below that purport to illustrate what happens to two, innocently-quiescent "***swatches***" of the "fabric of Spacetime" located in a ***sheet*** of such ***normal*** to the progression of a gravitational wave consisting of a ***pair*** of independent, ***complementary*** "***modes***" (my notation but see below). The pair share the ***quadrupole energy*** in the wave, and we are about to see the ***history*** of the swatches as they encounter the wave. The modes are 45 degrees apart and can be ***quiescent*** or ***extended*** to where they have accepted the full energy of the wave. Thus, each Mode may be in either of three "***Conditions***" (still my notation):

1. Fully ***Quiescent***
2. Fully ***extended***
3. In ***transition*** between the two conditions meaning that A) the extension can be growing at the expense of the quiescent condition or B) the reverse; quiescence grows while extension diminishes.

A couple of caveats: In the figures we don't show the ***transition*** condition. Also, the quiescent figure should be a ***circle*** and the extended figures should be long ***ovals***. However, I have drawn all shapes with straight lines which are much easier to do on the drawing

[2]Including famed Nobelist Richard Feynman who approached Gravitation as due to the existence of "Gravitons" (Fenyman, Morinigo and Wagner1995).

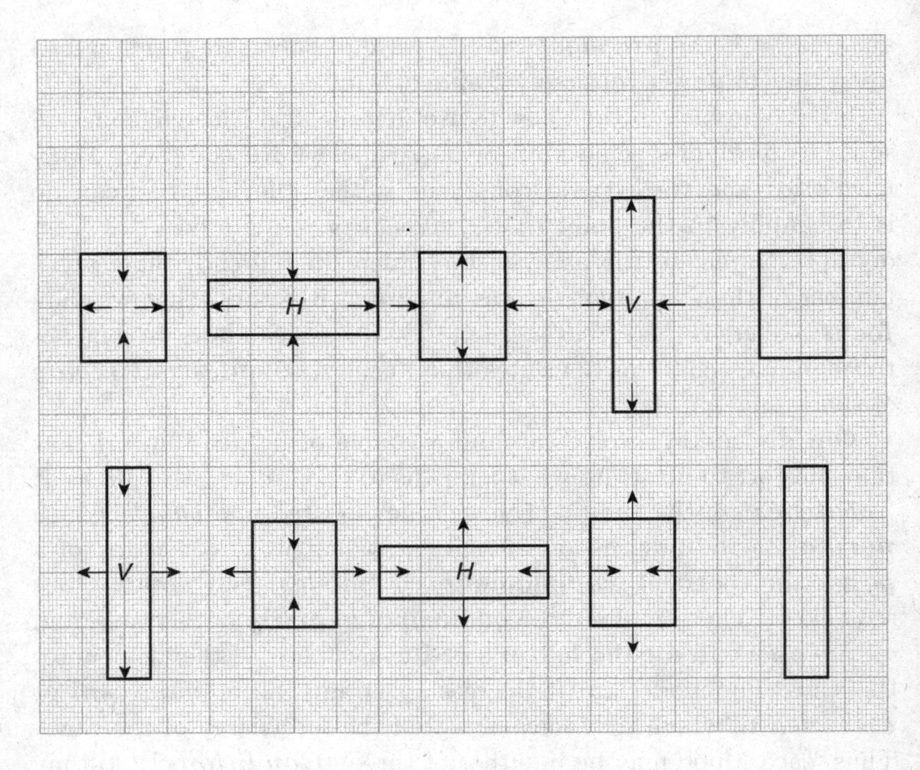

Fig. 18.1. Quiescent Sequences

board I saved from my drafting class in High school a hundred years or so ago! So the circle becomes our square, the extension of which can be in either of two ***complementary directions*** in swatch coordinates, ***Vertical*** or ***Horizontal***. As graphically illustrated by Feynman these two are the ***only modes*** compatible with the Spin 2, ***quadrupole*** nature of gravitational radiation. However he does not show the progression we present herewith; the main reference for which is.

In Fig. 18.1, the upper sequence shows a set of fully ***quiescent*** and fully ***extended*** conditions for the first mode and the lower sequence the same for the "slanted" mode but lagging by ***45 degrees***. We note that quiescence and extension are each ***90 degrees*** of the

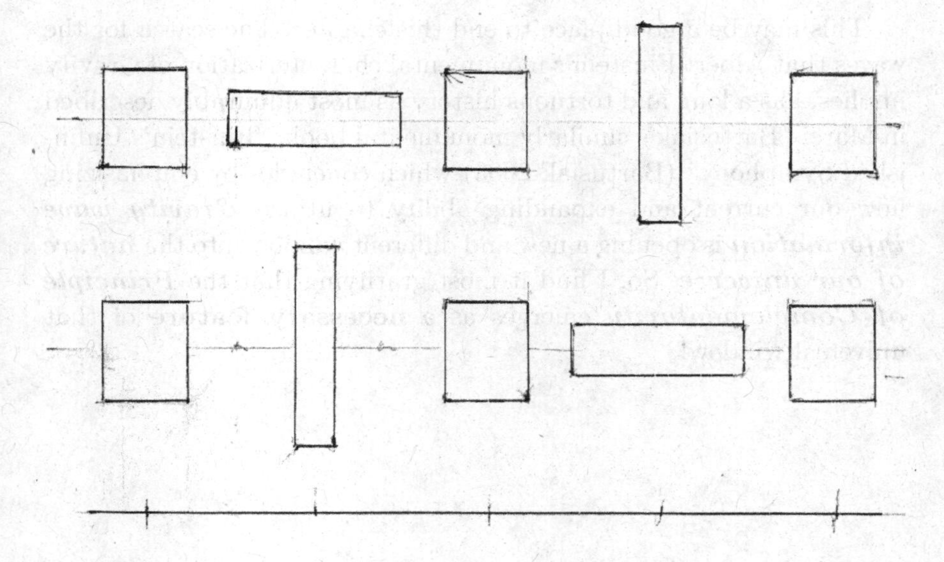

Fig. 18.2. Advanced Quiescent Sequences

gravitational wave apart but vertical and horizontal extension are each ***180 degrees*** apart which seems to make sense.

In Fig. 18.2, we have ***advanced*** the lower sequence by ***45 degrees***. As a result, we note that we now have the ***upper*** sequence of horizontal-vertical coinciding with the ***lower*** sequence of vertical-horizontal, something we can summarize as the ***2 × 2 matrix***

$$\begin{pmatrix} H & V \\ V & H \end{pmatrix}$$

which is readily re-expressed as our familiar ***Complementarity Signature matrix***,

$$\begin{pmatrix} 1 & -1 \\ -1 & 1 \end{pmatrix}!$$

In other words, a ***Gravity wave*** may also be viewed as a ***manifestation*** of ***Complementarity***!

This may be a good place to end this chapter. The search for the waves that Albert Einstein's monumental characterization of Gravity implies, has a long and tortuous history as most admirably described in Marcia Bartusiak's similarly monumental book, "Einstein's Unfinished Symphony" (Bartusiak 2003) which concludes by emphasizing how our current and expanding ability to utilize ***Gravity wave information*** is opening a new and different window into the ***nature of our universe***. So, I find it most gratifying that the ***Principle of Complementarity*** emerges as a **necessary feature** of that universal window!

19

Another look at Our "Indigenous Higgs Model"

In 1992, in a question — and — answer book intended to explain the highlights of Superstring Theory, the immensely knowledgeable physicist ***Edward Witten*** began his response to the question "What are the essential problems that the superstring theory claims to address?" thusly: "In twentieth century physics there are two really fundamental pillars, one of them is general relativity which is Einstein's theory of gravity and the other is quantum mechanics, which is the theory of everything that goes on in the microscopic domain. In other words it's the theory of atoms, molecules and smaller objects called elementary particles. The basic problem in modern physics is that these two pillars are incompatible (Witten 1992)."

We need not go into Witten's description of the disheartening things that occur in the mathematical formalism when it is attempted to conjoin the two theories in some way, shape or form. However, I do have a suggestion as to why that is the case and here it is:

The ***two theories*** simply *exist in two* ***disjoint categories***. As the Ancients used to say, "Fire and Water, Earth and Air, etc."!

Of course we know that gravity is "***real***"; it's a manifestation of the ***curvature of Spacetime*** as described by the left side of the famous Einstein equation, which, as Nobelist ***David Gross***, put it, also in the question-and-answer book, is that in effect, Einstein wanted to somehow move the right-hand side of the equation over to the left — in other words "to understand matter as a geometrical structure. To build matter itself from geometry — which, in a sense, is what string theory does (Gross 1992)".

Well, we shall not contrast the superstring approach to elementary particle physics with that of our Alternative Model at this point. The matter of interest here is that, in contrast to the "***reality***" of gravitational phenomena, what the ***formalism of QM*** provides us are the relative ***probabilities*** of a set of empirical outcomes rather than a definite outcome itself, the realization of which earned physicist ***Max Born*** a Nobel prize (Nobel Foundation 2021)!

Another, and perhaps somewhat more sophisticated, way of saying the same thing is that ***Einstein's theory*** is an ***ontological*** approach to the nature of ***Gravity*** whereas ***Quantum Mechanics*** constitutes an ***Epistemological*** approach to understanding the nature of very **small entities** and their interactions so, although they are ***Complementary***, there is no point in trying to ***meld them together*** into a single ***theoretical formalism***.

Nevertheless, how then are we able to construct our "***Indigenous Parallel***" to the ***Higgs*** mechanism, a formalism that operates within the confines of the elementary particles of our model and that includes three ***fundamental constants***, the speed of light, c, the gravitational constant, G and *Planck's* constant, h, without running into the problems cited by Witten.

Basically, the answer is two-fold: from an overall point of view, our approach is essentially ***Ontological*** rather than ***Epistemological*** and, in more detail, the ***formalism*** we construct ***does not*** really make use of Quantum Mechanics (QM), *per se*. Both Einstein's "Ansatz" that underlies our approach and Bohr's model of the proton atom are ***explicitly ontological***: nowhere do we see anything of a statistical nature that would be characteristic of QM. Our adaptation of Bohr's "***old Quantum Mechanical***" model was freely enabled to include Planck's constant. Then, we simply ***postulated*** that the resultant force emanating from the center of our ***Solitonic model*** is ***gravitational*** rather than electromagnetic under the admittedly ad hoc assumption that gravitational effects at those ***tiny scales*** are still a mystery!

In any event, the result turns out to be equitable to our so-called "NI Theorem" which is similarly related to the Higgs model. ***On the other hand,*** let's be honest: the words "***tiny scales***" should be a warning; perhaps we cannot legitimately avoid a ***quantum mechanical approach*** to the characterization of the gravitational force emanating from the center of our solitonic geometry!

Well, the next section is devoted to quantum mechanics so we can take a look at this situation therein.

IV

Quantum Mechanics and the Significance of Scale

20

Introducing Max Planck and the Quantum

The 20th century was an amazing epoch. My mother was born in January of 1900 and lived to the age of 95, something of great significance to me personally. Of more interest to this monograph and indeed to this chapter is a paper I downloaded from the Internet entitled "One Hundred years of Quantum Physics" by Daniel Kleppner and Roman Jackiw, both of the Massachusetts Institute of Technology (Kleppner and Jackiw 2000). The opening paragraphs rank, for me, among the most effective, informative such I've ever read, and I would like to quote them as follows:

"An informed list of the most profound scientific developments of the twentieth century is likely to include general relativity, quantum mechanics, big bang-cosmology, the unraveling of the genetic code, evolutionary biology, and perhaps a few other topics of the reader's choice. Among these, quantum mechanics is unique because of its profoundly radical quality. Quantum mechanics forced physicists to reshape their ideas of reality, to rethink the nature of things at the deepest level, to revise their concepts of position and speed, their notions of cause and effect."

"Although quantum mechanics was created to describe an abstract atomic world far from daily experience, its impact on our daily lives could hardly be greater. The spectacular advances in chemistry, biology and medicine — and in essentially every other science — could not have occurred without the tools that quantum mechanics made possible. Without quantum mechanics there would be no global economy, for the electronics revolution that brought us the computer age is a child of quantum mechanics, as is the photonics revolution that brought us the information age. The creation of quantum physics has transformed our world, bringing with it all the benefits — and the risks — of a scientific revolution."

To which I can only add "that says it all"! On the other hand: perhaps you recall the mention in the Preface of the book by Greenstein (G) and Zajonc (Z) (1997) and my characterization of it as "their penetrating analysis of the epistemological quandaries of quantum mechanics." In essence what G and Z were talking about is the problem of ***interpreting*** Quantum Mechanics (QM). They mention at some length the long-running debate between Einstein and Bohr and how Albert "— emphasized that the theory had relinquished precisely what had always been the goal of science — the discovery of the ***'real'***" (my emphasis) while Bohr's "— Copenhagen interpretation insisted that this tradition — had now to be abandoned".

Or, in more pithy, down-to-earth terms, there is also a reference in their book to Richard Feynman's assessment that "Nobody understands quantum mechanics", uttered well before what G and Z characterize as the enormous technological strides that have occurred in recent years. And that might have been expected to shed some light on the subject. But that "has only made the theory's nature more evident". So, in summary, the paradox is still with us; we still, in fact increasingly utilize the utility of QM while remaining, possibly even increasingly so, also puzzled by its nature!

Well, so much for epistemology and philosophy, at least for now; the point is that QM is loaded with that stuff, and in fact, if we think about it we're sort of back to "the nature of ultimate reality" and *The Meaning of "Is"*, subjects I thought I put to rest early on but we'll turn to again later in the book. But, historically speaking, QM sure didn't start out with any such concerns in the mind of the one individual credited with initiating the process that led to it, coincidently at the beginning of the century, in fact, in the year 1900 (I don't know how close it was to my mother's birthday). No, Professor ***Max Planck's*** main interests (other than music;

he had at one time considered becoming a musician!) were primarily in ***thermodynamics*** and he had no idea of the momentous nature of his part in the beginning of the process that led to all those mind-boggling developments mentioned above.

Nor did anyone else. It took five years before anyone published anything that had anything to do with what Planck came up with and sure enough it was that inherent revolutionary ***Albert Einstein***! Well, here's how it all transpired, at least the way I understand it.

In the first place, as far as physics was concerned nobody (including Planck) had ever heard of quantum mechanics or anything containing the word "***quantum***" which is derived from the Latin, although most European languages have similar words related to the idea of "amount". To my knowledge, that is. Anyway, what ***Planck*** was interested in, and intensively so at the time was a way to put together a mathematical expression matching the measured spectrum of "***blackbody***" ***electromagnetic*** radiation. It had been half a century since the Maxwellian revolution and physicists were quite familiar with the Electromagnetic (EM) radiation associated with a broad range of phenomena. In particular, they had made a lot of measurements of the energy spectrum of EM radiation emitted by a so-called "***blackbody***" with the temperature of the body as a parameter.

To them, "***blackbody***" basically meant an enclosure that absorbed all incident external radiation with a small hole through which the EM radiation reciprocally emitted to and received from the cavity by its walls could be sampled, radiation that was observed to come to a steady state at a given temperature. Figure 20.1 is a freehand representation of such a spectrum as a function of radiation ***frequency*** which is what physicists with a background in E and M kind of things would tend to choose as the independent variable. (Those with an optical background would of course tend to go with wavelength instead but one has to make a choice.)

We see a rising and falling structure with a peak that, it turns out, tends to shift over to the right with increasing temperature according

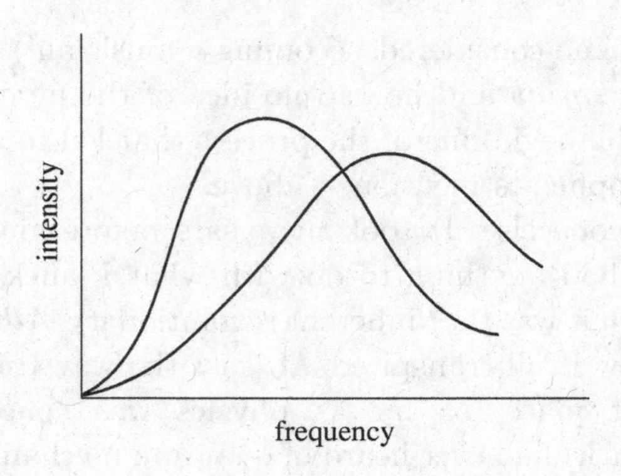

Fig. 20.1. Two Planck "blackbody" Spectra

to ***Wien's law*** of displacement[1];

$$f_m = 2.822kT \tag{20-1}$$

where T is temperature in degrees Kelvin and k is the Boltzmann constant that came up in Chap. 8 on Thermodynamics.

The problem Planck (and everyone else) had was to write down an expression that reproduced the ***entire*** spectrum of energy versus frequency observed at a given temperature. There were at that time mainly two such ***partial*** distributions: The ***Rayleigh-Jeans*** law starts at zero frequency and does all right for a while but then it diverges without limit with increasing frequency according to

$$dE/df = RkT \tag{20-2}$$

where $R = 16\pi^2 f^3$ and dE/df is energy per unit frequency interval and per unit enclosure volume. Meanwhile the ***Wien*** version (not the displacement law!) comes down from infinity with increasing frequency to make a pretty good fit on the RHS of the figure according to

$$dE/df = \mathrm{Re}^{-hf/kT}. \tag{20-3}$$

[1]This the first time that the ***augmented***, ***complementary*** combination, kT that fully expresses the energy equivalent of heat (as characterized by its temperature) has come up but it will again!

Given that situation, after some cogitation Planck had an "epiphany" and divined what amounted to an ***interpolation*** formula between the two extremes according to

$$dE/df = Rhf\mathrm{e}^{-hf/kT}/(1-\mathrm{e}^{-hf/kT})$$

$$\text{or} \qquad (20\text{-}4)$$

$$dE/df = Rhf/(\mathrm{e}^{hf/kT}-1)$$

your choice, which, to his immense satisfaction, duplicates the Raleigh-Jeans and Wien laws at low and high frequencies, respectively[2].

But then (I suspect) he got to thinking: What he had was his ***own*** idea, ***Planck's law***, and he should be able to derive it from first principles. Well, that task turned out to be a lot harder to accomplish. He struggled and struggled with it until he hit on a procedure that while successful was inexplicable to him at that stage in the history of physics. Instead of the differential format as in the above, what he did to begin with was to divide the radiation emitted by the walls into a ***discrete set of miniscule frequency intervals***, Δf, and assign to each such interval at frequency f a miniscule "***quantum***" of energy, hf. Then, treating the totality of such quanta as a "***gas***" of independent entities and using ***thermodynamic*** reasoning he proceeded to derive his unique distribution in the limit of vanishing interval size, as we see above.

All very gratifying but mysterious not just to Max but to everyone else and there the matter lay until ***Einstein's*** "miraculous year" of 1905 when among his projects at that time was something known as the "photoelectric effect" wherein monochromatic light impinging upon a metallic surface induced electrons to fly off it in a seemingly anomalous way as a function of the parameters of the light. While the ***number*** of electrons leaving increased with light intensity, their maximum ***energy*** did not, it instead increased in a

[2]And, finally, here we see the ratio of the ***quantum of energy***, at a given frequency, to the energy equivalent of heat (as per its temperature).

linear way with light ***frequency***. It occurred to young Albert that if Planck's radiation was indeed emitted in a discrete way by the walls of the blackbody chamber, perhaps that discreteness was also a characteristic of the ***radiation*** itself! In other words the light impinging upon the metallic surface in photoelectric experiments was ***itself*** composed of ***"quanta" of radiant energy***. Following up on that premise then led to the simple relationship for individual electron energy

$$E \leq hf - \Phi \tag{20-5}$$

where f is the frequency of the radiation and the constant Φ, known as the "work function" of the particular metal used in the experiment was attributed to internal forces that had to be overcome before the electron could be free to leave. It also led to a 1921 Nobel prize for Einstein but not before his relationship was experimentally verified (in 1916) by Robert ***Millikan*** (who also got a Nobel). (Why Einstein did not get one for his "Special" Relativity is a long story better left for another venue.)

Not long thereafter (in 1923), Arthur ***Compton*** contributed another experimental verification of the ***discrete*** nature of electromagnetic radiation at low levels. The "Compton effect" is exhibited when individual electrons are exposed to low level light to show that there is such a thing as a ***momentum*** to be attributed to individual "quanta" of light by then known as ***Photons***. In fact, what Compton found was that the relation of the wave length λ', of the light after collision to λ, that before collision could be expressed as

$$\lambda' = \lambda + \lambda_c(1 - \cos\theta) \tag{20-6}$$

where θ is the angle through which the incident photon is scattered by the collision and $\lambda_c = h/mc$ is the "Compton wavelength". A theoretical explanation of what transpires is readily concocted using straightforward classical mechanics that comply with the dictates of relativity with a little help from what Prince Louis-Victor de Broglie contributed a year later in his doctoral thesis. (There's no law in France that a member of the remnant nobility cannot get involved in the nature of physical phenomenology). That's when the Prince introduced his famous, simple but cataclysmic relationship between particle ***momentum*** and the associated ***wavelength***, $p = h/\lambda$. Of course, academic requirements being what they are there was a lot more of substance to the thesis than just that simple straightforward statement and it ended up being a most influential thesis (for example de Broglie was able to show that each "Bohrian" ***orbit*** must encompass a ***whole number*** of his royal waves).

In any event, using that statement with a little algebra (I know; they always say that but you might like to mess with it) shows the Compton relationship to be equivalent to the relativistic expression for the invariance of total energy before and after the collision of a massless particle of momentum p, namely a ***photon***, with a stationary particle of mass m, namely an ***electron***. My main reference here is a very lucid book by P.J.E Peebles (1992) otherwise known for his comprehensive cosmological contributions and here's the way he expresses it:

$$pc + mc^2 = p'c + \left[m^2c^4 + \left(\overrightarrow{p} - \overrightarrow{p'}\right)^2 c^2\right]^{1/2}. \tag{20-7}$$

Here the LHS is the total initial energy, a sum of that of the impinging photon plus the stationary electron derived from the relative

$$E^2 = p^2c^2 + m^2c^4 \tag{20-8}$$

while on the RHS $p'c$ is the energy of the scattered photon and $\left(\overrightarrow{p} - \overrightarrow{p'}\right)^2 c^2$ is the energy transferred to the electron.

According to Peebles a good way to begin is to subtract $p'c$ from both sides, square both sides and simplify to obtain

$$p' = \frac{mcp}{mc + p(1 - \cos\theta)}. \tag{20-9}$$

I tried it and it works! Not only that, if you introduce the de Broglie relation into this last equation, it gives you just what Compton found out as above!

At first glance, the Compton experiment is a most elegant, indirect way to find out something, in this case the transfer of momentum, by actually measuring something seemingly quite different, namely a change in wavelength. However, when you think about it in terms of the de Broglie relationship between particle momentum and wavelength, it's not indirect at all — a change in momentum is indeed equivalent to a change in the wavelength associated with the particle's wavelike nature. That would be starting to think quantum mechanically!

21

Quantum Mechanics, Phase 1

Well, there are some other important things that emerged about this time before the big quantum revolution of the mid-twenties that we'll get to shortly. These include ***Bose-Einstein*** and ***Fermi-Dirac*** statistics, the identification of ***spin***, ***Pauli*** exclusion and more, some of which well also get to later but for now, I want to sketch out what is perhaps the salient feature of this introductory phase, the ***Bohr model*** of the atom, something ***Niels Bohr*** put together (literally) in 1913. Here's a thumb nail version of what he did, taken mainly from what I put together in my previous book:

As we know, a hundred years or so ago, Bohr formulated a model of the Hydrogen atom that predicts that element's discrete line spectra. It had already been established (by **Rutherford** in 1911 and recognized with a Nobel Prize) that the known atoms consisted of a tiny, centrally located nucleus with a positive charge and negatively charged electrons whose disposition in the atom where still mysterious (APS 2006b). As a young man, ***Bohr a Dane*** had spent some time in Rutherford's Manchester laboratory and had become interested in the mystery himself. In characteristic fashion, he boldly decided to investigate the case of electrons ***circling about the nucleus*** even though it seemed obvious to many that they would eventually just end up there due to electromagnetic attraction. So he wrote down the attractive force,

$$F_E = e^2/r^2, \tag{21-1}$$

between a proton in the nucleus and an electron in orbit as balanced by the ***centrifugal force***

$$F_C = m_e v^2/r \tag{21-2}$$

experienced by the electron due to its ***orbital velocity***. Here, e, r, m, and v are electronic charge, orbital radius, mass and speed in orbit, respectively.

And here's where the "Bohrian" individuality comes into play: he simply ***postulated*** that angular ***momentum*** is ***quantized*** in multiples of Planck's constant according to

$$mvr_n = nh/2\pi. \tag{21-3}$$

Where n refers to the nth orbit in a supposedly concentric set of such. The combination of these relationships yields an expression for the ***radius*** *of that orbit*, namely

$$r_n = h^2 n^2/4\pi^2 me^2. \tag{21-4}$$

Whereupon (***complementarily***!) balancing ***Coulomb's law*** of attraction and ***centrifugal force*** it is also possible to compute the total (kinetic plus potential) energy of an electron in the that orbit as

$$E_n - e^2/r_n \tag{21-5}$$

which, combined with Eq. (21-4) for the radius gives the formula for the ***energy*** in the nth orbit as

$$E_n = \frac{-2\pi^2 me^4}{n^2 h^2}. \tag{21-6}$$

But Bohr had not used up all his daring: he also ***postulated*** that ***energy*** in an atom could only change when an electron changes its ***location*** by a whole ***number*** of ***orbits*** and that the energy of the atom thereby changes by a ***whole quantum***. In other words, dropping down from a higher to a lower orbit would be accompanied by a changed energy of amount

$$E_i - E_f = hv \tag{21-7}$$

and the emission of a ***photon*** whose frequency is v which, on the basis of the previous two equations can be expressed as

$$\nu = \frac{-2\pi^2 me^4}{n^2 h^3}\left(\frac{1}{n_f^2} - \frac{1}{n_i^2}\right). \tag{21-8}$$

To give this long story a climax, Bohr's theory was an instant success; his formula matched precisely the spectra experienced for the hydrogen atom down to the numerical value of the coefficient preceding the parenthesis above, which Bohr associated with the labeling of the orbits he had constructed so cleverly. By the way, going back to ***de Broglie*** once more, the Prince was able to show that each such Bohrian ***orbit*** must encompass a ***whole*** number of his royal waves. (And there's more; see below.)

Unfortunately, as Bohr himself knew, his theory was really too simple and its shortcomings were not long in showing up; little things like ***Degeneracy*** which, characteristically, reared its head as soon as more refined experimentation was developed. But here Bohr had a lot of help in the modeling. First, ***Arnold Sommerfeld***, a professor of physics introduced ***Special Relativity*** and ***elliptical*** orbits and when ***spin*** was recognized, ***Wolfgang Pauli*** took care of that as well as of the closely related effects of ***magnetism***. In fact, Pauli ended up putting together a complete model of the hydrogen atom that included three more quantum numbers which, in addition to Bohr's original, took into effect all of the above.

22

Phase 2: QM Through the Looking Glass!

Case closed, right? Well, sort of. Unfortunately, trying to fit that model to anything more complex than Hydrogen turned out to be infuriatingly difficult, in fact impossible, as was realized and acted upon primarily by four heroes of the quantum revolution over a span mainly of just a few years of the second decade of the twentieth century. The first was ***Werner Heisenberg*** at that time a researcher in a physics group headed up by Professor ***Max Born***, once a graduate student of ***Minkowski*** (whom we met earlier in connection with Relativity).

Actually, Heisenberg was Born's assistant; although only 23 at the time he had been recommended for the post by Sommerfeld his doctoral mentor. According to ***K.S. Lam*** (1992), Heisenberg was motivated by the notion[1] that "Physical ***theory*** ought to focus on quantities immediately related to ***observables***. Since the Bohr orbits used in the early quantum theory are never directly observable, they should not play a direct role in the theory. Realizing that all observable quantities such as emission frequencies, transition rates, etc. — are always related to a ***pair*** of Bohr ***orbits***, never just to a single one, Heisenberg sought to build a Quantum Mechanics (QM) based on theoretical constructs linking the totality of all such of a given atomic system".

Since for all practical purposes there would be an infinity of orbits — and here is the heart of his contribution — Heisenberg ***conjectured*** that for each basic ***observable*** there should correspond

[1]Which he may have formed in part by conversations with Einstein.

an infinite, ***two-dimensional array*** of measurable numbers. There is a well-known story about how, plagued by an allergic condition, Heisenberg decamped to an island in the North Sea where he worked all night with arrays of numbers representing known data to verify his conjecture until he was ready to show his associates what he felt was a novel approach to quantum mechanics general enough to apply to atoms more complex than Hydrogen. A real breakthrough!

The situation was that not just Heisenberg but Born's group in general had been preoccupied with the problem of extending the Bohr model. As Born put it, "In Göttingen we also took part in the attempts to distill the unknown mechanics of the atom out of experimental results. The logical difficulty became ever more acute — the art of guessing correct formulas — was brought to considerable perfection — ".

So Heisenberg's new perspective was for the most part welcomed by Born except that he couldn't quite grasp the mathematical significance of what he referred to as **quadratic arrays**, especially Heisenberg's discovery that ***two such arrays***, say those associated with the orbital parameters p and q, when multiplied together obeyed the rule,

$$qp - pq = ih/2\pi \tag{22-1}$$

where q and p are the canonical (and ***complementary***!) ***position*** and ***momentum*** variables of phase space. That is to say that unlike ordinary numbers, they ***did not commute***.

Which reminds me of another story; it won't take long and it has bearing on the main discussion. Many years ago... well, actually at least decades ago or so... I happened to share an elevator at work with a young friend of mine who I knew was not that many years away from having earned his doctorate in physics. What Bob (I don't remember his last name) was doing in a company involved with space and missile things is another story but not unusual. Anyway, I asked him a question about something I had been thinking about in my spare time, namely, what the connection was if any between classical physics and Quantum Mechanics. It took him some time too cogitate about it ... maybe 15 seconds or so before the elevator doors opened but his answer, delivered with some confidence, was "***Poisson Brackets***" objects, I dimly recalled from my own days as a physics major, that had something to do with classical mechanics at least the way it had been developed beyond the basic Newtonian model thereof.

Well, segue through the years to this book and what I had to do to prepare myself for writing this chapter and, lo and behold — Bob was absolutely right, and I'll talk about why that is in a minute! It has to do with ***Max Born*** who was probably the most qualified of all those exposed to the new model to understand its mathematical significance. After puzzling over it for a week he suddenly recalled ***his*** student days and his exposure to the mathematical objects called ***Matrices*** — Voila! — Heisenberg's arrays should be called ***Matrices***! Immediately he and ***Pascual Jordan*** a mathematically minded member of his crew went to work and with Heisenberg's collaboration they put together a definitive article describing Heisenberg's version of quantum mechanics. The ***Matrix Mechanics*** of fundamental particles was born (no pun intended)!

So, here's what Lam designates as Heisenberg's final equation of motion[2]:

$$\frac{du}{dt} = \frac{[u, H]}{i\hbar} + \frac{\partial u}{\partial t} \tag{22-2}$$

where u and H are "in general infinite matrices" as per the ***Born/ Jordan*** formalization and $[a, b] \equiv ab - ba$ is the ***commutator*** of matrices a and b.

In comparison to what we might regard as some basic mechanics, here's what Lam also calls the ***Hamiltonian*** equations of motion of an arbitrary function in ***phase space***

$$\frac{du}{dt} = \{u, H\} + \frac{\partial u}{\partial t} \tag{22-3}$$

[2]Some place I read that it was really Dirac who originally came up with that formalism.

where the curly bracket houses the ***Poisson bracket***

$$\{u, v\} \equiv \sum_i \left\{ \left(\frac{\partial u}{q^i} \right) \left(\frac{\partial v}{\partial p^i} \right) - \left(\frac{\partial u}{p^i} \right) \left(\frac{\partial v}{\partial q^i} \right) \right\} \tag{22-4}$$

which we met in Chap. 7 and which summarizes the effect of the ***Hamiltonian function*** on the variation of the function u in time. The point here is the basic **formal resemblance** that exists between the ***bracket*** and the ***commutator,*** something instantly recognized by ***Dirac***.

Before we go on with our story, there is another point that, it seems to me ought to stressed, and that is the structural resemblance of ***both*** equations not only to ***each other*** but to the ***covariant derivative*** wherein either the ***commutator*** or the ***Poisson bracket*** plays the role of the ***connection.*** The implication is that the Heisenberg approach to quantum mechanics constitutes a ***gauge theory*** featuring ***curvature*** in some way, shape or form, something we talked about in Chap. 12.

In the meantime, we note that Heisenberg's getting together with Born and Jordan was not the end of the story; he was very fortunate in having someone like Max Born to tell his story to, especially someone who turned out to have the background needed to frame the story in the proper mathematical context, the ***matrix*** formulation. Unfortunately, familiarity with matrix algebra was not widespread and not everyone was convinced. In particular, off in another corner of the world of deep thinkers, Professor Erwin ***Schrödinger*** was digesting the novel point of view espoused in ***de Broglie's*** doctoral thesis. In particular, he was intrigued by the de Broglie ***relationship*** between ***momentum*** and ***wavelength*** $p = h/\lambda$ and his thesis that, just as what are generally considered to be **waves** should also exhibit "***particle-like***" behavior, what are generally viewed as ***particles*** should also exhibit "***wave-like***" behavior!

In fact, ***Schrödinger*** was so taken aback with de B's thesis that he gave a colloquium on it in which he emphasized the implications for ***particle*** behavior. His talk was attended by a number of other influential physicists one of whom pointed out that it was all very interesting but something important was missing, namely an explicit ***wave equation***! At which point Schrödinger decided he had better devote his considerable talents to putting together just that: a **wave equation** describing the behavior of ***particles*** like electrons, in general and especially in the context of their membership in atoms. In that regard he was very much impressed with ***Hamilton's*** recognition of the ***analogous formalisms*** of classical ***mechanics*** and ***geometrical optics*** and, as noted by ***Lam***, he ended up going one step further by establishing a relationship between the ***wave mechanics*** of particles and ***physical optics***.

You may recall wave equations having been discussed elsewhere in the book; they generally have the structure

$$\frac{dy^2}{dx^2} = \left(\frac{1}{v^2}\right)\left(\frac{dy^2}{dt^2}\right). \tag{22-5}$$

Well, that's not exactly what Schrödinger came up with! It took some time and several publications to discuss all the ramifications of his work in that regard but here is his most salient result, the famous ***Schrödinger Equation***, strenuously exercised by several generations of physicists (and chemists) to reveal the detailed quantum mechanical behavior of all manner of physical phenomena:

$$\left(-\frac{\hbar^2}{2m}\nabla^2 + V\left(\overrightarrow{r}\right)\right)\psi = E\psi \tag{22-6}$$

where $\hbar = h/2\pi$, E and V are the total (constant) ***energy*** and ***potential energy***, respectively, the latter a function of position and the expression in the large parenthesis being the Hamiltonian, H so that we often see expressed simply as

$$H\psi = E\psi. \tag{22-7}$$

As noted above Schrödinger was initially motivated by *de Broglie's* novel contribution and although it's not immediately obvious, that's reflected in his equation. To see that, we note to begin with, that for an electron in free space (actually, it need not be an electron;

it could be another particle) Eq. (22-6) becomes

$$-\frac{\hbar^2}{2m}\nabla^2\psi = E\psi, \qquad (22\text{-}8)$$

a solution of which in the form

$$\psi \sim e^{-i\vec{k}.\vec{r}} \qquad (22\text{-}9)$$

then leads to

$$\nabla^2\psi = -k^2\psi, \qquad (22\text{-}10)$$

if, that is, $E = \frac{\hbar^2}{2m}k^2$. We also note that, for free electrons, $k = 2\pi/\lambda$ which implies that $E = \frac{\hbar^2}{2m}\left(\frac{2\pi}{\lambda}\right)^2$. Finally, if we now recall the ***de Broglie relation*** $\lambda = h/p$ we end up with $E = p^2/2m$, one of the foundations of mechanics. In other words, at least in free space, the de Broglie relation is seen to be a necessary ingredient in the formulation of the Schrödinger Equation.

I should interject a most important point here about the ***interpretation*** of Quantum Mechanics: Differential equations being familiar to most physicists, Schrödinger's formulation (in contrast to the Heisenberg theory) was an instant success. Except that no one really understood what it was that was being differentiated!

So, here's where ***Born*** stepped in once more to save the day, this time earning himself a Nobel prize: What Max explained was that $\psi\psi^*$ (that is not ψ itself but the square of its absolute value obtained by multiplying it by its complex conjugate) was to be regarded as the ***probability*** of some physical quantity. Thus, did ***statistical uncertainty*** emerge as an integral element of particle physics.

Interjection:

In my opinion, this is where the matter of ***scale*** emerges as significant; what I'm saying is that when the things we're concerned with are small enough, we might find it necessary to talk about them in statistical terms. Anyway, I'll talk about it some more in the next chapter.

Well, at this point I'm going to terminate this rather sketchy discussion of the background of quantum mechanics. The ***Bohr***, ***Heisenberg*** and ***Schrödinger*** theories, along with ***Born's*** contributions I should think, are the most important to this early part of the story even though ***Dirac's*** contributions to unification should really not be omitted, especially in terms of the ***uncertainty*** principle. For example, ***Lam*** (See References) uses the Dirac ***notation*** to develop it beginning with the ***commutation*** relationships (which implies ***Poisson Brackets***) but I shall forgo that development hoping to minimize distracting complexity at this point. Later on, other formulations such as Richard ***Feynman's*** summation over all path's become most useful but that's another story.

As far as ***complementarity*** is concerned ***Bohr's*** original assessment tells it all; the ***Heisenberg/Born matrix*** formalism being concerned with particles in terms of their **corpuscular** nature and the ***Schrödinger equation*** in terms of their associated ***undulatory*** behavior ... particles or waves ... what is by now the common wisdom is what ***Bohr*** said which was that what you ***see*** depends on the kind of ***experimentation*** you ***choose.***

However, there is one ***completely different*** approach to QM that I cannot ignore because it helps to highlight a basic aspect of how quantum mechanics relates to the rest of physics and indeed to our view of the world we inhabit. Not only that but it makes ***uncertainty*** automatically manifest as the minimization of a ***position-momentum*** area in ***phase space*** rather than as the ***separate*** minimization of the ***product*** of an extension in position and an extension in momentum. It's called the ***Wigner Function*** and it's the subject of the next chapter.

23

Quantum Mechanics, Radar and The Significance of Scale

This Chapter has a special meaning for me; for one thing, in a way it represents another example (see Chap. 4) of the identification, seemingly, of two quite different sets of phenomenological considerations. As per the title of this chapter, one of the two is radar, my association with which began almost 70 (I think) years ago! Actually it began with a combination of radar and missile guidance because the company I had started working for was in the missile business, specifically the development of a "beam-rider" surface-to-air missile. What that means is that the missile continually corrects its location in the cross section of the beam transmitted by a radar that tracks a target the missile is supposed to intercept and demolish — old-time technology!

Anyway, my first job was to build and test a "breadboard" of what they called a roll computer designed to keep the missile from going into an uncontrolled roll. For starters I didn't know what a breadboard was, and I wasn't terribly familiar with circuit diagrams nor with the value of the various electrical components that I was supposed to attach to the breadboard. After all I had been a physics major and had spent the previous couple of years getting an MS in physics to the neglect of anything practical!

But I learned; one of the main things being to avoid getting too involved with breadboards and such and more into the theoretical and systems analysis side of the business. However, another important item that gradually seeped into my consciousness was that there seemed to be a decided resemblance between Heisenberg's ***Uncertainty principle*** in Quantum Mechanics (QM) and what was known in radar parlance as the "***Radar ambiguity function***".

So now, segue to after I had retired and gotten involved in Particle Physics and Knot theory. I was at a symposium that, if I recall correctly, was dedicated to the memory of the acclaimed Nobelist ***Eugene Wigner*** and got to chatting with one of the key speakers, Thomas L. ***Curtright***, a Professor of Physics at the university of Miami (see below) who spoke about Wigner's approach to QM and when I mentioned my observation, he concurred. Actually, it occurred to me that there seemed to be a similarity between the Wigner approach and radar altogether, but in any event, armed with that encouragement I should have pursued the subject but, regrettably, I let it become submerged in the affairs of the day.

Now, however, it would seem that this book has given me a perfect opportunity — perhaps a duty (!) — to take it up once more and, as we shall see, demonstrating that formal similarity in terms of ***complementarity*** turns out to be quite straightforward. That's on the plus side; on the minus side is the fact that nowadays, from what I see on the internet, that formal similarity to a class of sensors of which Radar is a prime example is fairly common knowledge. Nevertheless my interest got a bit of a boost when I happened to see Curtright's name as the coauthor of a book (Curtright, Fairlie and Zachos 2014) that ***features*** the ***Wigner Function*** approach to QM based on viewing it as a "quasi-probability" distribution function in phase space of position x and momentum p,

$$f(x,p) = \frac{1}{2\pi}\int dy\psi\cdot\left(x-\frac{\hbar}{2}y\right)e^{-iyp}\psi\left(x+\frac{\hbar}{2}y\right), \tag{23-1}$$

namely as "— a generating function for all special autocorrelation functions of a given quantum mechanical wave function $\psi(x) = \langle x|\psi\rangle$".

You will recall, I trust, that ***location***, x (or y) and ***momentum***, p, are, ***manifestly complementary***, as conjugate variables in ***phase space***, that celebrated region of dynamics and Quantum

Mechanics, and subject to that celebrated restriction, the ***Heisenberg Uncertainty Principle***. As Curtright, and etc. emphasize, the ***Wigner function*** approach " — furnishes a third, ***alternative***, formulation of quantum mechanics, independent of the conventional **Hilbert space** or path integral formulations. In this logically complete and self-standing formulation, one need not choose sides between coordinate or momentum space. It works in full phase space, accommodating the uncertainty principle and it offers unique insights into the classical limit of quantum theory: the variables (observables) are *c*-number functions in phase space instead of operators, with the same interpretation as their classical counterparts, but composed together in novel algebraic ways."

Although we never got into it in the preceding, what Curtright & Co. were alluding to was that while the Hilbert space approach to QM ***encompasses*** the formalization of ***both*** the Heisenberg and Schrödinger approaches as well as the path integral approach that Richard Feynman devised, their claim is that the Wigner function approach is a valid **third** approach that offers some unique insights — and what I like, in particular, — "into the ***classical limit*** of Quantum Theory".

Which brings us back to Radar and (please wait for it) the ***Radar Ambiguity*** Function! Well, to begin with, radar's early days (in the early days of WWII) were a frantic, hit or miss struggle to put together something that would detect invading aircraft soon enough to intercept them in the crucial "Battle of Britain". Since then, Radar has matured into a sophisticated art as well as a science with a large diversity of applications. Nowadays we have, of course, Sonar for acoustic applications, Lidar for the use in the optical domain, etc. and a lot of sophistical developments we'll just touch upon below.

By the way, as you probably know, Radar is an acronym for "Radio Detection and Ranging" where "ranging" means continual estimation of target distance ("range") accomplished by measuring the time delay of target backscatter (its ***echo***) of the transmitted signal back to the radar receiver. For our purposes here I shall confine our deliberations mainly to the case of the detection and tracking of a single target.

Here's a rather top-level account of what happens at the Radar: First a "***carrier***" waveform is generated at the radar frequency which, typically for tracking, might be in the range of several hundred up to perhaps tens of thousands of megacycles, depending on antenna limitations, atmospherics, target characteristics, etc. Then, since we need that carrier to carry some information, a "marker", whose echo will tell us something about target range at any instant, we need to apply some kind of ***modulating*** signal to the carrier which, for all practical purposes, means multiplying the carrier by the modulation.

Here, the sky's the limit; the modulation can be continuous or pulsed, it can involve the variation of amplitude or of phase or frequency, according to either a simple, repetitive format or a more complex, mathematically-generated scheme — you name it. We can also get information about the target's velocity toward or away from the radar by the rate of change of range but a faster and more accurate way is to directly measure the so-called ***Doppler frequency*** offset of the target's echo which is instantaneously proportional to both range rate and carrier frequency (see below). That requires that the Radar be ***"coherent"*** meaning that it can detect the ***phase*** as well as the ***amplitude*** of the received signal.

And finally, the modulated carrier is sent through an antenna that shapes the transmission's angular occupation in space, or more specifically the so-called far-field cross-section of the illumination otherwise known as the ***antenna pattern*** (or ***beamshape***, basically the characteristics of a cross-section of the beam). Depending on various implementation factors, beamshape, as measured at, say, its half-power locus, is also subject to a lot of variation but, typically, this may vary from a circle to a long, very slender ellipse known as a fan beam (that I'll talk about later on for the case of a specific application other than target tracking). But, you may ask, how can we keep the target in the beam? No problem; for instance, just arrange for the receiving antenna to generate a split pattern — two overlapping beams — and compare the signals received in each to obtain an error signal that we put through a tracking loop. I'm oversimplifying of course but no matter; that's not really important to our main thesis here.

What is important is the aforementioned ***ambiguity function*** which can be expressed in terms of the transmitted signal $u(t)$ as

$$\theta(\tau, \phi) = \int_{-\infty}^{\infty} dt u(t) u^*(t + \tau) e^{-i2\pi\phi t}, \tag{23-2}$$

where the conjugate, ***complementary*** variables in this case are, respectively,

$\tau =$ the ***difference*** between R_T/c, the out-and-back ***delay*** time of the radar signal modulation envelope (due to target *location in real space*), and R_M/c, the time to which the ***processing*** system is matched, R being target range and

$\phi =$ the ***difference*** between $V_T f_0/c$, the Doppler frequency ***offset*** (due to target velocity V_T) of the radar carrier waveform and $V_M f_0/c$, the Doppler frequency offset to which the ***processing*** system is matched, which, if we identify

$$\begin{aligned} \tau &= 2\left(R_T - R_M\right)/c \\ \phi &= 2\left(V_T - V_M\right) f_0/c \end{aligned} \tag{23-3}$$

where f_0 is the carrier frequency and c is the speed of light.

It took a while but now that we've got that out of the way, I note that what we would really like to see is the equivalent ***symmetrical form***

$$\theta(\tau, \phi) = \int dt\, u\left(t - \frac{\tau}{2}\right) u^*\left(t + \frac{\tau}{2}\right) e^{-i2\pi\phi t} \tag{23-4}$$

that we can more readily compare with the ***Wigner function*** which we repeat here:

$$f(x, p) = \frac{1}{2\pi} \int dy\, \psi^*\left(x - \frac{\hbar}{2} y\right) \psi\left(x + \frac{\hbar}{2} y\right) e^{-iyp}. \tag{23-5}$$

Despite the difference in nomenclature the formal ***structural*** resemblance between the two is quite obvious; we see each with an exponential term with a complex, two-term, ***complementary*** exponent as well as the product of two, complex, binomial functions of complex arguments. Nevertheless, we can make it a bit tighter by a few simple, arbitrary transformations on the Wigner function, viz:

1. We start out by equating the exponents in the two expressions since they refer to similar spaces, ***complementary*** differential ***delay*** and ***Doppler*** frequency offset in the radar case and position/momentum in the Wigner case; basically, location and velocity in both cases. That is, we set
$$yp = 2\pi\phi t. \tag{23-6}$$
2. We invoke de Broglie, for whom momentum is given by $p = h/\lambda$ with the quantum of action.
3. Finally, we ***arbitrarily*** equate the wavelength λ to the inverse of the Doppler offset — i.e. according to
$$\lambda = c/\phi. \tag{23-7}$$

Putting 1, 2 and 3 together, we find that the quantity in parenthesis in (23-4) becomes

$$\hbar y = \tau \tag{23-8}$$

whereupon we have

$$dy = 2\pi(c/h)dt \tag{23-9}$$

and

$$f(x,p) = (c/h)\int dt\psi^*(x - ct/2)\psi(x + ct/2)e^{-i2\pi f_D t}. \tag{23-10}$$

Finally (!) making the identifications

$$x \leftrightarrow t \text{ and}$$
$$\psi\sqrt{c/h} \leftrightarrow u^* \tag{23-11}$$

the Radar and Wigner formulations become identical; check it out!

If this *ad hoc* procedure seems rather arbitrary, let me try something else on you, namely (to drop a famous name once more); how about Max Planck's "***interpolation***" formula we talked about in Chap. 20, the one that introduced, for the first time in the history of the world, the quantum, h? If it worked for Max to find the correct

Blackbody radiation expression at ***low*** frequencies, the ***Rayleigh-Jeans*** law and the correct expression at ***high*** frequencies, the ***Wien function***, why can't it work for us?

Well, let's take another look at the two formulas we compared above and see what's involved in comparing what seems like such widely diverse phenomena; remember DNA and Particle physics? Here, on the one hand, the reason Radar and quantum mechanics can be ***similarly formalized*** is that they are really concerned with similar ***situations***. That is, they are instrumented to help us say something about the kinematics and dynamics of our world in ways that, given the sensorial limitations we are born with, we cannot.

Their ***difference*** arises mainly in the obvious circumstance that their focus is vastly different in ***scale***. Radar behavior must be taken into account when we use it to look at our ***external*** world in the range of from, say, ***meters*** to ***astronomical*** distances. Contrastingly, as we go to ***smaller and smaller*** scales, eventually our knowledge becomes, ***inherently,*** only ***statistical***, whereupon, as per the ***Born interpretation*** (probabilistic) and as discovered by Planck (Chap. 20) to begin with, dynamics and electrodynamics must yield to their Quantum Mechanical counterparts wherein, for the first (and only) time, h, the ***quantum*** of ***Action***, emerges and in fact becomes ubiquitous.

So, what does the Planck formula do? Here it is, reproduced from the above:

$$dE/df = Rhf\{e^{-hf/kT}/(1 - e^{-hf/kT})\}. \tag{23-12}$$

As Planck realized, at low frequencies we can approximate this as

$$dE/df = Rhf\{1/(1 - [1 - hf/kT])\} \tag{23-13}$$

and at high frequencies as

$$dE/df = Re^{-hf/kT}. \tag{23-14}$$

So, to change the ***radar ambiguity*** function to a ***Wigner-like*** expression, both the ***delay***, τ and ***Doppler difference***, ϕ entries in that function must become very small when delay and Doppler difference ***themselves*** are small. In other words we want to change them as they appear in the ambiguity function so as to feel comfortable at that scale.

First, we can modify the function τ to read

$$\tau/2 \to \tau/2\left[\frac{1}{1-e^{-(\tau/\hbar y)}}\right] \tag{23-15}$$

which, at small values of delay, τ becomes

$$\tau/2 \to \tau/2\left[\frac{1}{1-(\tau/\hbar y)}\right] \approx \hbar y/2 \tag{23-16}$$

as in the Wigner function.

However, at large values the exponential function disappears, and we have the original, $\tau/2$.

Similarly, we modify the exponent $i2\pi\phi t$ to read

$$i2\pi\phi t\left\{\frac{1}{1-e^{-(i2\pi\phi t/yp)}}\right\} \tag{23-17}$$

which becomes $2\pi\phi t$ at small values of ϕ, as in Wigner, but remains the same at large values.

That's it! I considered letting everyone make those modifications on the original Radar Ambiguity Function (RAF) for themselves but then I thought that my diligent and astute readers deserve better than that so here's what the modified RAF looks like a Planck:

$$\theta(\tau,\phi) = \int dtu\left(t - \frac{\tau/2\left[\frac{1}{1-e^{(-\tau/\hbar y)}}\right]}{2}\right)u^*\left(t + \frac{\tau/2\left[\frac{1}{1-e^{-(\tau/\hbar y)}}\right]}{2}\right)$$

$$\times \exp(i2\pi\phi t)\left\{\frac{1}{1-\exp(-i2\pi\phi t/yp)}\right\}. \tag{23-18}$$

Which, as per the above, does, indeed produce the ***Wigner*** function for very ***small*** values of both τ and ϕ and the RAF for ***large*** values, something that hopefully gives us a way to rationalize how QM occurs as the right way to systemize nature as the scale of interest is reduced from the domain of Radar utility down to the micro-and-submicroscopic scales.

So much for the Wigner/ Radar comparison; now back to some more about the radar by itself: Some of you may have noticed that we glossed over the fact that Radar performance is also ***highly statistical*** but, at least superficially, for different reasons than QM. The signal emitted by the Radar is tasked with the duty to go out into the world and ***convolve*** with ***all*** the scatters therein, ***not only*** the ones deemed desirable by the Radar's users (But see below). As a result, depending on many factors, geometrical, physical, application specific, etc. the signal returns loaded with ubiquitous, undesirable signal known as "***Clutter***" generally characterized only by its statistics.

Which is a major problem, one to be dealt with by signal design and processing. However, it's not the only problem: Upon its return the signal is also contaminated by an insidious additional component that has nothing to do with target echo. It's called **noise** and, without getting into detail, suffice it to say that noise can originate ***within*** the radar system itself as well as ***outside*** from a number of sources usually but not always expected and at least statistically describable. (Most of us know about the unheralded Radio Astronomers who won a Nobel prize by being unable to eliminate the low-level, low-temperature, broadband noise whose statistics seemed invariant to where in the sky they pointed their antenna designed to scan the heavens. And how analysis of that annoying affliction led ultimately to the apparent corroboration of the theory of the "Big Bang" origin of the Universe!)

Anyway, the early radars were strictly non-coherent (no phase information) but, eventually, stable ***oscillators*** and ***phase-lock*** systems were developed, and it became possible to do ***coherent*** signal processing (see above). In Radar parlance it became possible to do ***both*** Range and Doppler processing and the main item of concern became how to obtain good resolution in ***both*** domains for a given transmitted waveform. That is, either (or both) how to reliably ***separate*** two targets or to ***detect*** a given target against a background of noise and clutter.

Now, as stated above, the ***time delay*** of the envelope of the return from a target provides information regarding its range (distance) while the ***Doppler shift*** of its carrier translates into velocity of approach. But clearly, good precision in time ***delay*** requires an envelope with very ***small extent*** in time, a delta function-like characteristic in the limit. Unfortunately, such waveforms are associated with very large Fourier spectral ***bandwidths***. Contrastingly, high precision ***Doppler*** frequency information requires signals that last ***long enough*** to encompass some number (at least one!) of cycles of the smallest frequency difference desired for Doppler resolution.

In other words, we would like to have ***both*** a large enough ***signal duration*** for good Doppler resolution and a large enough ***bandwidth*** for good range resolution and a measure of that combination is the so-called ***Time-Bandwidth*** product. So signal ***duration*** and its Fourier transform ***bandwidth*** are thus ***complementary*** quantities and, a given waveform is associated with a particular ***Time-Bandwidth product*** and its distribution ***in Range-Doppler*** space from which we can deduce the signal's resolution capabilities. Which brings us back to ***Quantum Mechanics***! We've pretty much established that Radar and QM are ***formally*** equivalent and now, clearly, the Radar ***Time-Bandwidth* product *in Range-Doppler space*** is simply the counterpart to the ***location-momentum product in QM phase space***.

And in ***both*** cases, it is ***because*** and ***only because*** we are involved with ***coherent processes*** and their ***Fourier transforms*** that this situation obtains. Which is what I tried to say back in the Preface; there is ***nothing mysterious***, nothing spiritual or philosophical about this situation; it's simply good, old-fashioned electromagnetics, geometry and kinematics and quite straightforward mathematics. Case closed (but see below).

By the way, I'm sure you've noticed something important, namely, why does the word "***ambiguity***" sneak into the discussion? Well, that output signal we wrote down basically expresses the output of the radar receiver for the case of a target signal unaccompanied by interference of any kind, ideally a narrow ***peak*** rising above some threshold signaling detection. In real life the interfering noise and clutter can provide false indications and the output is ***ambiguous*** to one extent or another depending upon the ratio of target return power to interference power.

Well, there's not much more we need to cover in this chapter. However, in terms of ***coherent signal*** processing there is a particular application I'd like to talk about, namely ***mapping*** Radar also known as Synthetic Aperture Radar (SAR). This is a ***side-looking*** Radar on an aircraft or a spacecraft using a ***fan beam*** that looks out and down in a direction normal to the platform's velocity. The fan beam is typically generated by a linear array of dipoles illuminating a long slender half-parabolic reflector on the side of the platform to generate a fan beam that, in moving along with the platform, scans the ground (actually, nowadays it's probably arrays of solid state dipoles integrated into the skin of the platform itself). Basically, the antenna pattern ***convolves*** with the ***scatterers*** on the ground; in other words, ground ***clutter*** has become the ***signal***!

Taking geometry and kinematics into account, we see that range information develops along the beam's intersection with the ground on each and every emission (typically a continual pulse train) while Doppler information develops with time parallel to the platform's track. The number of pulses available for Doppler processing depends on the platform's velocity, the pulse repetition rate, and, because the fan beam intersects the ground in a basically triangular illumination pattern, the range of interest and the angular width of the fan. The resultant processing proceeds coherently across all the pulses available in the illumination pattern to derive Doppler information within the enhanced integration time. And, we note, the processor in the air or spaceborne receiver processes all the pulses received during the associated temporal interval and we have the "***synthesis***" of an enhanced aperture.

Now, consider a "patch" of terrain, small relative to the area illuminated at any one time but large enough to have "structure", that is, it itself is made up of component scatterers. As the beam sweeps by, the patch doesn't move but the Radar does and therefore the line of sight to the patch changes its ***angular orientation*** from pulse to pulse — it ***rotates*** — making a different "cut" through it upon each pulse.

We note that this ***rotation*** of the line of sight also underlies "***Tomography***" which is the basis for the familiar Computer

Assisted (CAT) scans used in the medical field to visualize areas of concern. So given the proper processing (what's known as "***matched filtering***" is often "optimal" in some sense) at the Radar receiver, that's exactly how our SAR maps the ground!

How about that? Here's another similarity between two, seemingly, ***widely diverse*** phenomenology, maybe not as striking as our earlier comparison between the particle world and DNA but gratifying nonetheless. Actually, there are various diverse applications of this basic line-of-sight rotation. For example, it is possible to let a reflecting body undergoing various motions about its center of mass to, in effect, enable its 3D picture to be taken by a coherent processing sensor!

Before we end this diversion, one last detail: how to derive adequate range information on a pulse-by-pulse manner presents a problem. The problem is solved in an ingenious manner; the frequency of the signal modulating the carrier frequency is slewed over the requisite ***bandwidth*** within ***each*** transmitted pulse. Upon reception, an ***identical*** modulation signal is used to demodulate each received pulse, thus regaining the initial signal amplitude within a narrow pulse shape with the desired range resolution. In other words, we have implemented a "synthetic aperture" in the carrier frequency domain so as to enable the desired narrow pulse width; our synthetic aperture is two-dimensional!

Well, that's all for our foray into ***Radar***; we began with its similarity to ***Quantum Mechanics*** despite the vast difference in scale and we've just ended by identifying it with a seemingly much different device employed in a completely different field and at an ***intermediate scale***, important to us personally and so justifiable. Along the way we put together a Planckian type interpolation method to relate Quantum Mechanics and Radar. All very gratifying.

So, in conclusion, what ***is*** the ***"significance of scale"***? Why put it into the title if we're not going to talk about it? Well, said significance is ***not*** in the ***mathematics***; complex mathematics (phase as well as amplitude) must be employed in both cases. And it is ***not*** in the need for a ***probabilistic*** point of view; both domains are plagued by random "interference" of one kind or another. We talked about that a bit in terms of radar phenomenology and as far as the quantum world is concerned, not only is it well-known to be plagued by inherent fluctuations but the Born Interpretation formalizes the need to take that into account.

The fundamental reason scale is significant is that somehow we find ourselves in the same situation Max Planck encountered 116 years ago; ***somehow*** the information of concern must be quantized in terms of the quantum of action, h, as very small values of action are approached. Otherwise the mathematics doesn't make sense in the real world just as Planck found out. So, the dilemma remains; is the world ***inherently*** so constructed or is it our need to fit our mathematical capabilities to it that is the problem? Stay tuned.

24

The Fourier Transform and the Convolution Theorem

As we saw in the foregoing, convolution and its partner the Fourier Transform are fundamental ways to understand what's going on in both Radar (as an example of a sensorial system) and quantum mechanics. I don't know for sure, but I think the Fourier Transform (FT) is better known to students just getting used to the ideas involved than is the idea of convolution. Historically, the FT provides one with the occupancy in a mathematical frequency space of things in real space, but it is often moot which space is the more real! Here's one way to express the FT; in one dimension we have

$$F(v) = \int f(v) \exp(2\pi i v x) dx \tag{24-1}$$

where the integral is over all x and we'll talk about it a bit more in terms of convolution below.

Convolution has a long history, but I first learned about it in relation to the response of a control system to input data or disturbance signals. That was a long, long time ago. I realized that the signal emitted by a radar, for example, is not just ***multiplied*** by the return from a target space; it ***convolves*** with it. I also found out that there's even a connection to ***probability*** theory associated with such phenomena. The literature on such matters discuss the sum of independent variables with mutually exclusive outcomes whose distribution is a convolution. (It's a little more complicated than this barebones summation but that's the general idea.)

Nowadays, ***convolution*** is featured in the processing of data sequences derived from a host of phenomena but the basic nature of a convolution is really simple if, that is, the variable underlying the sequences in question is ***discrete*** (see below). For example, the convolution of two such sequences means just taking their ***sum*** as they ***slide past*** each other. You may recall how we used a version of that kind of thing to develop a taxonomy of the elementary particles!

However, if the variable is continuous that summation must be formally expressed as an integral, viz:

$$C(h) = \int f(x)g(h-x)dx = \int g(x)f(h-x)dx. \tag{24-2}$$

Notice that, in effect, f and g are ***complementary*** contributors to the formation of c and it doesn't matter which does the sliding!

You may recall how we sort of took this formalism in vain when, in Chap. 3, we combined two ***discrete*** sequences of fermionic ***labels*** (not numbers!) to group the degeneracies of each of the values of boson twist that occur in the 4×4 boson matrix. The 4-vector of fermions and the 4-vector of antifermions are also ***complementary*** contributors to this grouping.

And you may also recognize that the operation of ***matrix multiplication*** is actually a ***convolution***. That is we consider both the rows and the columns of a square matrix as vectors when, in multiplying, say, matrix A on the left and matrix B on the right we perform an inner multiplication of each of the columns of B by each row of A as those rows of A move successively past the columns of B. For example, suppose we have the matrix product

$$\begin{bmatrix} a & b \\ c & d \end{bmatrix} \begin{bmatrix} e & f \\ g & h \end{bmatrix} = \begin{bmatrix} ae+bg \\ ce+dg \end{bmatrix} \begin{bmatrix} af+bh \\ cf+dh \end{bmatrix} \tag{24-3}$$

then, first vector (a,b) multiplies vector (e,g). Then it goes on to multiply vector (f,h) while vector (c,d) is busy multiplying vector (e,g). Finally, vector (c,d) also multiplies vector (f,h).

So, we can think of one "super" vector sliding past the other and computing along the way by multiplication of the overlap and summing, viz:

$$\{(cd)(ab)\} \Rightarrow \{(eg)(fh)\} \tag{24-4}$$

Finally, here's a very useful and important relationship between ***convolution*** and Fourier theory: it's called the ***Convolution Theorem*** and it's expressed in two ***complementary*** ways, viz:

1. The FT of a ***convolution*** of two functions is the ***product*** of their ***individual*** FTs.
2. The FT of a ***product*** of two functions is the ***convolution*** of their ***individual*** FTs.

That's neat, don't you think? And the reason the relationship is important is almost obvious: the processing involved in either case is thereby made a lot simpler. Here's a proof of the first theorem, starting out with the convolution

$$C(h) = \int f(x)g(h-x)dx \tag{24-5}$$

as per the above. Its transform (sorry for the change in notation) is then

$$\begin{aligned}
F(v) &= \int C(v)\exp(2\pi ivx)dx \\
&= \int \left\{ \int f(y)g(v-y)\exp(2\pi ivx) \right\} dydx \\
&= \int \left\{ \int f(y)g(u)\exp(2\pi ivx)\exp[2\pi i(y+u)]dy \right\} dx \\
&= \int f(x)\exp(2\pi ivx)dx \int g(u)\exp(2\pi ivu)du \\
&= \int f(x)\exp(2\pi ivx)dx \int g(x)\exp(2\pi ix)dx
\end{aligned} \tag{24-6}$$

which looks formidable but we can break it down into the following steps:

1. The first line is just the FT of that product of two convolutions.
2. In the second line we just change variable $(v - y)$ to u which is compensated by changing variable v in the second exponent to $y + u$.

3. This enables the ***crucial*** change in the third line from the double integral to the product of two independent integrals after which,
4. Line four is just a trivial change in notation.

So there you have it; the ***product*** of two Fourier Transforms as the ***transform*** of the ***product*** of two ***convolutions***. Nothing could be simpler, right? And the second theorem is more or less the same kind of progression.

By the way, the theorem is useful in signal processing where the FT of the system's ***Impulse Response*** is it's ***Frequency Response***. (Actually we're usually dealing with the Laplace Transform but its the same kind of situation.)

25

Spin, Spinors, and the Pauli Connection

We begin this chapter with a quotation: "No one fully understands spinors. Their algebra is formally understood but their general significance is mysterious. In some sense they describe the 'square root' of geometry and, just as understanding the square root of -1 took centuries, the same might be true of spinors." The speaker, a mathematician of highest repute in many areas including the fields of particle physics and its relationships, shall be nameless, mainly because I seem to have mislaid the reference for this rather gloomy prognostication. Were he to become aware of it, I should hope he would forgive me for saying so, but I think we may be able to do better than that; I don't think I can afford to wait centuries and perhaps the rest of this chapter will indicate why we don't really need to.

I recognize, of course, that the general subject of spin and spinors is indeed complex and that various types of spinors with some celebrated names attached to each have been defined to explicate various physical and mathematical situations. However that's not our concern here. Our concern is mainly with the elementary particles of our Alternate Model and, in that regard, the characterization of those particles given in the preceding bears repeating: "they are to be regarded not as discrete, point-like objects in a vacuum, nor as quanta of a field but, rather, as ***localizable-sustainable distortions*** in and of an otherwise featureless continuum. Indeed, they are not seen as '***objects***' at all in the ordinary sense of the word, but as ***topological entities*** — ***solitons*** — which persist because they are *topologically non-trivial*; they cannot dwindle away to a point and disappear."

From this point of view, Alternative Model particles are ***classical*** (*not* ***quantum***) entities. Nevertheless, a most important, intrinsic consequence of the modeling (see below) is that they ***manifest*** the fundamental ***quantum*** *attribute of* ***spin***. In fact, they constitute ***manifestations*** of ***spinors***. And as remarked in the preface to a textbook on the subject "— the foundations of the concepts of spinors are groups; ***spinors appear as representations of groups***." Which implies that ***our particles*** constitute ***manifestations*** of ***groups***, or more explicitly, it turns out, a ***particular group***, namely the gauge group SU(2). So, all things considered, at least for our purposes, there is no problem understanding spin or spinors!

But, let's talk about it a bit anyway; it's an interesting situation. In the Quantum Mechanical model of the elementary particles, the concept of the attribute of spin was originally postulated in 1925 by ***Goudsmit*** and ***Uhlenbeck*** in order to explain the fine structure observed in atomic spectra in the presence of a magnetic field and was then formally elucidated in terms of the Pauli/Dirac theory. A basic element of this concept is the theory of ***Spinors***, entities which were actually introduced by **Cartan** in 1913 (Tod 2006), well before they made their way into particle physics. In our model, however, as per the above, they emerge as ***classically describable*** entities inherent to the ***toroidal ontology*** of the MS itself.

SU(2), the group of special, unitary, 2×2 complex matrices, is well-known to map onto the group SO(3) of rotations in 3-space in a two-to-one manner, that is, such that a rotation by angle θ, expressed as an SU(2) matrix operation on the 2×2, self-adjoint matrix

$$X = x_k \sigma^k = \begin{bmatrix} z & x - \mathrm{i}y \\ x + \mathrm{i}y & -z \end{bmatrix}, \tag{25-1}$$

noted by ***Cartan*** to be ***equivalent*** to a net rotation by 2θ of the ***vector*** $\vec{x} = (x, y, z)^{\mathrm{T}}$ in 3-space. Here the σ^k are the ubiquitous **Pauli** (should really be Cartan/Pauli) spin matrices (see below) which, together, constitute a three-component vector whose inner product with the x^k is Eq. (25-1). This circumstance is customarily used in explicating the unusual nature of spin which, although it is measured in units of angular momentum, occurs ***quantized in all half-integer*** rather than only in ***integer*** multiples of $h/2\pi$ where h is Plank's constant. We saw a number of informal (mainly graphical) manifestations of this very circumstance in Section II in terms of the basic concepts and development of our particle model.

Interjection: We note that the Pauli matrices,

$$\sigma^1 = \begin{pmatrix} 0 & 1 \\ 1 & 0 \end{pmatrix}, \quad \sigma^2 = \begin{pmatrix} 0 & \mathrm{i} \\ -\mathrm{i} & 0 \end{pmatrix}, \quad \sigma^3 = \begin{pmatrix} 1 & 0 \\ 0 & -1 \end{pmatrix}, \tag{25-2}$$

occupy a pivotal position in the development of a variety of mathematical formalisms central to physical theories. As per the characterization above they are indeed ubiquitous. Not only that but we see them all over the place. One might go so far as to characterize them as ***central to the entire subject of spin, spinors and the associated group SU(2)***. For one thing, it is readily demonstrated that they constitute a basis for the Lie algebra associated with SU(2). However, there's more to their claim to fame, some of which we discuss below.

Also, we recall, the development of our ***taxonomy*** was systematized to begin with in terms of the group SU(2). The reason for the choice of SU(2) was stated (per Wigner) as being because it is the so-called "little group" associated with the orbits of particles whose kinematics are located in the (relativistically accessible) "forward light cone" of spacetime, with energy-momentum vectors

$$p = (p_0,\ p_1,\ p_2,\ p_3)^{\mathrm{T}}$$

such that $p_0^2 > 0$ and

$$m^2 = p_0^2 - (p_1^2 + p_2^2 + p_3^2) > 0,$$

where m is particle mass, and all its irreducible representations are parameterized by spin in multiples of 1/2. Here, the matrix-equivalent to the Lorentz 4-vector, is formed by substituting p_1, p_2, and p_3 for x, y, and z, respectively, in Eq. (25-1), and adding p_0 to z.

All of which is mainly of academic interest only by way of justification for the group theoretic summary of taxonomical development and we will not be considering kinematics in our model. However, Eq. (25-3) ***is*** of direct interest to our model because, in the first place, as per Cartan, it constitutes a ***spin matrix,*** whose elements define the fundamental two component ***spinor***

$$\xi_0 = \sqrt{(x - \mathrm{i}y)/2}$$
$$\xi_1 = \sqrt{(x + \mathrm{i}y)/2}. \tag{25-3}$$

Note that the diagonal (that is, the z) elements of the spin matrix are not involved in this formalism, which implies that spin is an essentially ***planar*** phenomenon. At the same time we recall, our concern here is with the notion of the MS as a ***concatenation*** of torus ***knots***, each of which, for visualization purposes, can be ***thought of*** as a real string wound around a real torus. The Cartesian coordinates of a point on such a string are (and here we're looking ahead a bit to the differential geometry of Section V)

$$x = w\cos\phi$$
$$y = w\sin\phi, \tag{25-4}$$
$$z = r\sin\theta$$

where $w = R + r\cos\theta$ is the projection in the x, y plane of the radius vector from the origin of coordinates to the point (ϕ, θ) on the toroidal surface, R is the radius of the toroid's circular centerline, r is the radius of its circular cross section, and ϕ and θ are angular measures in the azimuthal and meridional directions, respectively. Then we can translate the spin matrix of Eq. (25-1) as

$$X = \begin{bmatrix} r\sin\theta & w(\cos\phi - \mathrm{i}\sin\phi) \\ w(\cos\phi + \mathrm{i}\sin\phi) & -r\sin\theta \end{bmatrix} \tag{25-5}$$

and the associated ***spinor*** components as

$$
\begin{aligned}
\xi_0 &= \sqrt{w(\cos\phi - \mathrm{i}\sin\phi)/2} \\
&= \sqrt{(w/2)\mathrm{e}^{-\mathrm{i}\phi}} = e^{-\mathrm{i}\phi/2}\sqrt{(w/2)} \\
\xi_0 &= \sqrt{w(\cos\phi - \mathrm{i}\sin\phi)/2} \\
&= \sqrt{(w/2)\mathrm{e}^{+\mathrm{i}\phi}} = e^{+\mathrm{i}\phi/2}\sqrt{(w/2)}.
\end{aligned}
\qquad (25\text{-}6)
$$

Thus, noting that spin has to do only with circulation in the xy plane, we see that the half-angle exponent on the right-hand side of Eq. (25-6) implies that a nominal traverse of 2π around the MS results in an increment of only π in the phase of the spinor components. In other words, we repeat, the MS provides a visualizable ***manifestation*** of the notion of a ***spinor*** (which, of course, we had already discussed in Section II but not quite in terms of this formalism).

Note, however, that the correspondence is not quite complete being complicated by the presence of the term involving $w = R + r\cos\theta$. Basically, this introduces what may be described as a "modulation" of the projection of the MS in the x, y plane as a function of ϕ. To see this a little more explicitly, note that the equations for the position vector components written out in detail are

$$
\begin{aligned}
x &= R\cos\phi + r\cos\theta\cos\phi \\
y &= R\sin\phi + r\cos\theta\sin\phi \\
z &= r\sin\theta.
\end{aligned}
\qquad (25\text{-}7)
$$

Clearly, x, y, and z are all periodic functions of ϕ. However, while the variation of z is just a single sinusoid, we can view how x and y vary in the sense of communication system theory as the amplitude modulation of a carrier signal $R\cos\phi$ by the modulating function $(1 + \rho\cos\phi)$ where $\rho = r/R$. That is, we would write

$$x = R(1 + \rho\cos\phi)\cos\phi \qquad (25\text{-}8)$$

and similarly for y with $\sin\phi$ instead of $\cos\phi$. To make the correspondence to a spectrum of frequencies, we can write

$$\phi(\ell) = 2\pi m\ell/L \qquad (25\text{-}9)$$

where ℓ is the index of ϕ values, $1 \leq \ell \leq L$, such that $\phi(L) = 2\pi m$. Consequently, the equation for x can be rewritten as

$$x = R(1 + \rho \cos 2\pi f_n \ell) \cos 2\pi f_m \ell \tag{25-10}$$

where $f_m = m/L$ and $f_n = n/L$.

The spectrum implied here is well-known to consist of a carrier signal at frequency, f_m, and a pair of sidebands at $f_+ = f_m + f_n$ and $f_- = f_m - f_n$, a fact readily demonstrated by rewriting the above expression in terms of a trigonometric identity as

$$x = R \cos 2\pi f_m \ell + r/2(\cos f_+ \ell + \cos f_- \ell). \tag{25-11}$$

Note that the sideband contribution continues to dwindle down as r/R goes to zero. However, the basic topological nature of the MS, we might say its ***character***, does ***not***. In other words, the toroidal winding of the MS boundary is ***essential*** to the separate ***identification*** of the individual MS regardless of the ratio r/R. Actually, we can be a bit more specific; we can identify particle ***spin*** with the ***carrier*** wave circulating around the particle in the plane and ***Isospin*** with the ***modulation***!

New subject: you may recall the "interjection" above where we mentioned the widespread use of the Pauli spin matrices. A most important instance of that is in the development of the **Dirac** theory of the electron which we outline in the next chapter where we shall see how the ability of the Dirac equation,

$$-DD^*\psi = (\gamma^\mu \gamma^v \partial_\mu \partial_v + m^2)\psi = 0, \tag{25-12}$$

to constitute a manifestation of the **Klein-Gordon** equation depends on the 4×4 gamma matrices satisfying the requirements of a ***Clifford*** algebra, namely that

$$(\gamma^0)^2 = I^4, (\gamma^i)^2 = -I, \text{ for } i = 1, 2, 3, \text{ and } \gamma^\mu \gamma^v + \gamma^v \gamma^\mu. \tag{25-13}$$

Where I^4 is the unit 4×4 matrix. A standard realization of this requirement is

$$\gamma^0 = (\sigma^0 \otimes I) = \begin{pmatrix} I & 0 \\ 0 & I \end{pmatrix} \quad \text{and}$$

$$\gamma^\mu = i(\sigma^2 \otimes \sigma^\mu) = i \begin{pmatrix} 0 & \sigma^\mu \\ -\sigma^\mu & 0 \end{pmatrix} \tag{25-14}$$

for $\mu = 1, 2, 3$, and the sigmas are the Pauli Matrices. But since it is readily verified (by explicit multiplication) that

$$\sigma^\mu \sigma^v + \sigma^v \sigma^\mu = 0, \tag{25-15}$$

t follows that

$$\gamma^\mu \gamma^v + \gamma^v \gamma^\mu = 0 \tag{25-16}$$

as well[1]. In a similar way, demonstrating that the $(\gamma^i)^2 = I$, for $i = 1, 2, 3$, is also straightforward so we see that the **Pauli** matrix algebra **guarantees** the validity of the **Clifford** algebra. Furthermore, in Chap. 5 we investigate the algebra of the ***Quaternions*** which, it turns out, is really a ***special case*** of the Clifford algebra. Thus the Pauli matrices also guarantee the validity of ***Quaternion algebra***.

And, by the way, both Eqs. (25-15) and (25-16) exhibit ***complementarity!***

We now show how the fundamental matrices associated with DNA and our set of four basic Fermions can be expressed in terms of Pauli matrices. First the latter are rewritten here with the addition of the unit matrix:

$$\sigma^0 = \begin{pmatrix} 1 & 0 \\ 0 & 1 \end{pmatrix}, \quad \sigma^1 = \begin{pmatrix} 0 & 1 \\ 1 & 1 \end{pmatrix}, \quad \sigma^2 = \begin{pmatrix} 0 & i \\ -i & 0 \end{pmatrix}, \quad \sigma^3 = \begin{pmatrix} 1 & 0 \\ 0 & -1 \end{pmatrix}. \tag{25-17}$$

Then we also copy matrix

$$m = \begin{pmatrix} 1 & -1 \\ -1 & 1 \end{pmatrix}, \tag{25-18}$$

and the three equivalent matrices introduced by Professor Avrin back in his suggestion of possible alternatives to matrix m,

$$\begin{bmatrix} 2 & 0 \\ 0 & 0 \end{bmatrix}, \begin{bmatrix} 0 & 0 \\ 0 & 2 \end{bmatrix}, \quad \text{and} \quad \begin{bmatrix} 2 & 0 \\ 0 & 2 \end{bmatrix}. \tag{25-19}$$

---

[1]We note for future interest that the gammas ***anticommute***.

Although it is straightforward to produce the required equivalences via a bit of algebra, it is not necessary because they are quite obvious by inspection viz:

$$\begin{bmatrix} 2 & 0 \\ 0 & 0 \end{bmatrix} = (\sigma^0 + \sigma^3)/2$$

$$\begin{bmatrix} 0 & 0 \\ 0 & 2 \end{bmatrix} = (\sigma^1 - \sigma^3)/2, \text{ and that} \tag{25-20}$$

$$\begin{bmatrix} 2 & 0 \\ 0 & 2 \end{bmatrix} = 2\sigma^0.$$

And, of course, we have that matrix

$$m = \sigma^0 - \sigma^1. \tag{25-21}$$

26

Nonlocality, Entanglement, and Complementarity

As a species, we have a concept of time as well as of space and to one extent or another, as the current manifestation of the species, we run our lives accordingly, even many of our scientific pursuits. However, with Einstein, Minkowski and Lorentz, we have graduated to a more sophisticated concept, the ***Relativistic*** concept of ***Spacetime*** that led to fundamental advances in Physics. Now, however, we are faced with a relatively new but related concept, the empirical notion of ***nonlocality***, a phenomenon that apparently depends on a requirement for what is known as ***entanglement***. Actually, rather than "now" I should really say "still" because the ideas involved have been with us and debated for on the order of eighty years and are still evoking fresh discussion!

In any event, what do we mean by "nonlocality" and "entanglement"? Well, here's an example that apparently has nothing to do with fundamental physics but every once in a while we see an article in the newspaper about it. It concerns identical twins, separated at birth and never in physical contact thereafter who apparently lead very similar lives — similar occupations, choice of mates, clothes, automobiles, etc. And you can't get much more "entangled" at birth than are identical twins.

Maybe the similarities are just apparent or statistically possible, but I thought I'd just mention it to get the juices flowing, as it were. So now back to basic physics and nonlocality. As discussed in the literature from an empirical point of view (the result of actual experimentation; but see the caveat below) this seems to be a quantum mechanical phenomenon of spatially separated pairs of elementary particles ***if*** they have been close enough to begin with, such that their physical descriptions are somehow closely related. In other words, if the particles are "***entangled***" to begin with, they remain so after separation regardless of their physical separation.

And, paradoxically, it was ***Albert Einstein*** who was in good part responsible for the original interest in nonlocality. I say "paradoxically" because even though he had been one of Quantum Mechanics (QM) founders, Prince Albert had a lot of trouble reconciling his understanding of the way the world worked, as we know, something he was able to employ to tremendous advantage for much of his life. As QM developed, he was bothered especially with the need to consider it in a statistical manner but also because of the gnawing feeling that it was somehow "incomplete"; that is, that there were unknown (and thus not taken into consideration) "***Hidden Variables***" of some kind or another that actually made the experimentation work.

He was not alone in this "feeling"; in 1927, the Nobelist L. De Broglie brought out his thoughts on the matter in terms of what he called a particle's "pilot wave" that could explain some experimentation but was terribly discouraged by the reception he received and desisted from further work at least for some time; a coterie of "experts" can be brutal — even when completely mistaken! However, David Bohm's later version which is more fully developed and in consonance with QM is still around and kicking to the best of my knowledge and, I understand, got L de B working again.

In any event, Einstein's position at the Institute at Princeton eventually endowed him with the help of some top-notch analysts, a grouping that was able to carry on physical analyses at a considerable philosophical level, much to his liking. Probably the most well-known of those endeavors was the publication in 1935 of a paper that suggested an experiment, the so-called EPR experiment named after the collaboration between Einstein, Podolsky and Rosen, the latter two being a pair of his analytical aids.

According to E. S. Fry, who by the way, was in involved in an important one of its ensuing experimental realizations, the idea of an EPR experiment was expected by the initiators to validate the argument that QM was not a "complete" theory but that it could become one by the introduction of a proper set of "hidden variables", somewhat of a misnomer, I think, for influences on a system that are not included in its basic quantum mechanical description.

Fry also describes a particular EPR experiment that I shall discuss later on in some detail, an experiment in which two ***photons***, emitted an ***atomic cascade*** and moving in opposite directions and whose polarizations are measured independently. The measurements can then be "predicted with certainty to coincide" by viewing them as "real" objects, without benefit of quantum mechanical considerations. (Actually, all experimentation to date has been conducted on photons!)

On the other hand, there is now Bell's theorem propounded by J.S. Bell quite some time later (and considered by many to be one of the most important such in modern history) and that, according to Wikipedia says, in its simplest form, "No physical theory of local hidden variables can ever reproduce all the predictions of quantum mechanics." However, Wikipedia adds that "the door is still open for ***non-local*** hidden variables." (emphasis added).

And, Wikipedia adds, Bell's conclusion that "In a theory in which parameters are added to quantum mechanics to determine the results of individual measurements, without changing the statistical predictions, there must be a ***mechanism*** whereby the setting of one measuring device can influence the reading of another instrument however remote. Moreover the signal involved must propagate instantaneously so that such a theory would not be Lorentz invariant." (Emphasis added).

Evidently a number of physicists have taken Bell's conclusion to heart because there are some major efforts going on to find such a mechanism, a couple of which are discussed in Scientific American. One such is authored by Juan Maldacena (Maldacena and Susskind 2013), well-known for a number of widely quoted fundamental analyses, who is pursuing the notion of Bell's "mechanism" being provided by a "Wormhole" in Spacetime that provides such conveyance from a Black Hole entrance station and a White Holeout exit station! A concept initially suggested by the team of Einstein and Nathan Rosen on relativistic grounds.

Another article authored by Scientific Editor Clara Moskowitz (2017) describes the pursuit of the notion that Spacetime itself may be "made of tiny building blocks of information" the binding together of which by initial quantum entanglement might enable nonlocal behavior.

That's all I'm going to say about current activity. At this point I want to go back to the kind of EPR experiment with ***photons*** described by Fry as per the above. However, I shall insist upon photons as modeled in this Book — that is as the ***first-order fusion*** of an ***electron and a positron***.

Then, with e for the electron and p for the positron, we now consider, in ***lines 1 and 2*** below ***two examples*** of the spatial disposition of recently emergent electron-positron pairs that provide ***maximal aversion*** for each other. In each case, we expect the recently emergent photons in the column on the left to flee toward the left and those on the right to flee toward the right. Also, we note that the photon on the left and the one on the right are ***entangled*** in each case by definition.

The arrows in ***line 3*** just represent the ***electric charge*** increment in switch attention from the electron to the positron (e to p). Finally, in ***line 4***, we represent an arrow to the right by a + and one to the left by a −, which we note, provides us with what we have defined in previous publication as the **signature** of ***Complementarity***.

$$[\underline{\mathrm{ep}} \quad \underline{\mathrm{pe}}]$$

$$[\underline{\mathrm{pe}} \quad \underline{\mathrm{ep}}]$$

$$\begin{bmatrix} \rightarrow & \leftarrow \\ \leftarrow & \rightarrow \end{bmatrix}$$

$$\begin{bmatrix} + & - \\ - & + \end{bmatrix}$$

In summary, what we have shown is that the initial photon disposition prior to separation, that is their ***entanglement***, is simply ***equivalent*** to the fact they are a ***complementarity pair***; in fact, that they constitute a pair of ***identical twins***, as it were, and must remain as such ***forever*** with all the attributes the twins possess, unless separated by an unforeseen circumstance!

It may be difficult to get used to the idea but that's all we need to know in order to accept the concept of ***nonlocality*** as exhibited by paired photons in the experimentations conducted to date. Nevertheless, as per Fry, if one photon passes through a particularly-oriented polarizer in a nonlocality experiment, the other photon must do so as well with complete certainty. One may characterize such photons as being equipped with the ultimate "Hidden variable" so that, given the inviolability of Bell's theorem, the implication is that they are simply ***not quantum objects***.

Of course the model of a photon as the fusion of an electron and a positron, in the first place, may take some getting used to as well, so it may help to consider Dirac's demonstration of the existence of electrons and positrons.

27

PAM Dirac, Mixmaster Extraordinaire

The preface to a small but most illuminating monograph on the life and work of Dirac (Pais, Jacob, Olive and Atiyah 1998) begins "Paul Adrien Maurice (PAM) Dirac was one of the founders of quantum mechanics and the author of many of its most important developments. He is numbered alongside Newton, Maxwell, Einstein and Rutherford as one of the greatest physicists of all time." And, we might add, his equation, universally known as the Dirac equation marks a revolutionary milestone in the story of Quantum Mechanics (QM) for several remarkable reasons; not only does it unite QM with Special Relativity, but in the process it introduces the ***inevitable*** existence of ***antiparticles*** and illuminates the fundamental particle attribute of spin.

Here, then, is an abbreviated version of what's involved, a version that, by the way, has been shown in several venues as well as in the referenced book:

As it is well known, the equation for a particle with rest mass m and spin 1/2 is succinctly stated by the Dirac equation

$$D\psi = 0 \tag{27-1}$$

where ψ is the quantum mechanical state vector, D is the Dirac operator given by

$$D = \mathrm{i}\gamma^{\mu}\partial_{\mu} - m, \tag{27-2}$$

and summation over $\mu = 0$ to 3 is implied with 0, and 1, 2 and 3 corresponding to the time t, and the spatial variables x, y and z, respectively. Also, a factor of $\hbar = h/2\pi$ in the first term has been set equal to 1. Dirac imposed a relativistic compatibility constraint by demanding that

$$-DD^*\psi = (\gamma^\mu\gamma^v\partial_\mu\partial_v + m^2)\psi = 0 \tag{27-3}$$

be equal to the Klein-Gordon equation,

$$(\partial^\mu\partial_\mu + m^2)\psi = 0, \tag{27-4}$$

the quantum mechanical equivalent of the Lorentz invariant

$$E^2 - p^2 = m^2 \tag{27-5}$$

with E and p being the energy and momentum associated with relativistic bosons and

$$D^* = (-\mathrm{i}\gamma^\mu\partial_\mu - m) = -(\mathrm{i}\gamma^\mu\partial_\mu + m) \tag{27-6}$$

the complex conjugate to the expression in Eq. (27-1).

All of this is of course well-known as is the resulting requirement that for Eq. (27-3) to be true the gammas must, as Dirac discovered, be constant 4×4 matrices that conform to the definition of a ***Clifford algebra***. This leads (***almost!***) directly to the re-expression of the Dirac equation as two, ***coupled***, vector eigenvalue equations, namely

$$\begin{aligned} \partial_t\psi_\alpha + S\psi_\beta &= m\psi_\alpha \\ \partial_t\psi_\beta + S\psi_\alpha &= -m\psi_\beta \end{aligned} \tag{27-7}$$

where $\psi_\alpha = (\psi_1, \psi_2)^{\mathrm{T}}$, $\psi_\beta = (\psi_3, \psi_4)^{\mathrm{T}}$ and

$$S = \begin{pmatrix} \mathrm{i}\partial_{\mathrm{z}} & (\mathrm{i}\partial_{\mathrm{x}} + \partial_{\mathrm{y}}) \\ (\mathrm{i}\partial_{\mathrm{x}} - \partial_y) & -\mathrm{i}\partial_{\mathrm{z}} \end{pmatrix} \tag{27-8}$$

which has traveling wave solutions with exponents proportional to linear combinations of the time and space variables — i.e. to $\omega t \pm \vec{k}\cdot\vec{x}$ where $\vec{k} = (k_x, k_y, k_z)^{\mathrm{T}}$, ω and $\vec{k}$ being the radial frequency and momentum eigenvector, respectively, thus implying the ***equivalent form***

$$\begin{aligned} \omega\psi_\alpha + K\psi_\beta &= m\psi_\alpha \\ \omega\psi_\beta + K\psi_\alpha &= -m\psi_\beta, \end{aligned} \tag{27-9}$$

where

$$K = \begin{pmatrix} k_z & (k_x - \mathrm{i}k_y) \\ (k_x + \mathrm{i}k_y) & -k_z \end{pmatrix}.$$

Either of these last two equations expresses the operation of what amounts to a 4×4 ***spin matrix***, on a ***four vector*** to generate the direct sum of two ***coupled spinors***, one with ***positive*** and one with ***negative*** energy. Note that the two interact with each other in a circularly referential, we might add, a ***Complementary*** fashion and, in that regard, they express the ***relationship*** alluded to above between the Dirac theory and our ***alternative particle*** model; that is they are talking about the same kind of situation. To see that, we begin by rewriting Eq. (27-9) as

$$\begin{aligned} \psi_\beta &= -K^{-1}(\omega - m)\ \psi_\alpha \\ \psi_\alpha &= -K^{-1}(\omega + m)\psi_\beta. \end{aligned} \tag{27-10}$$

We can then demonstrate the ***circular*** nature of this pair of equations explicitly by the operational diagram of Fig. 27.1, below, where the operators are the annotated boxes.

However, we can also display the exactly identical situation in another way, as in Fig. 27.2: What we see there is a ***topology***

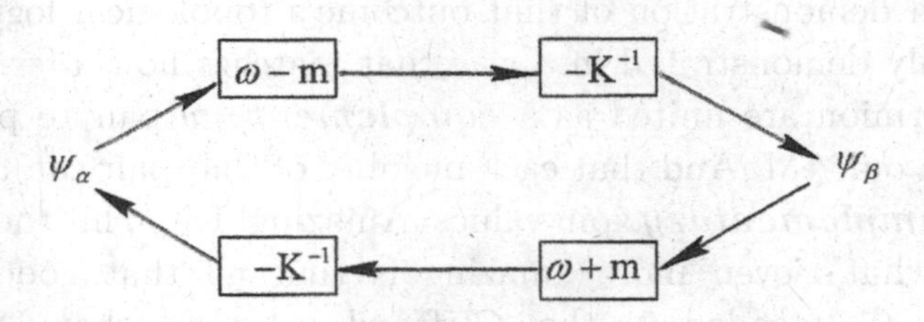

Fig. 27.1. A figure that highlights the "Circularity" of the Dirac Equations

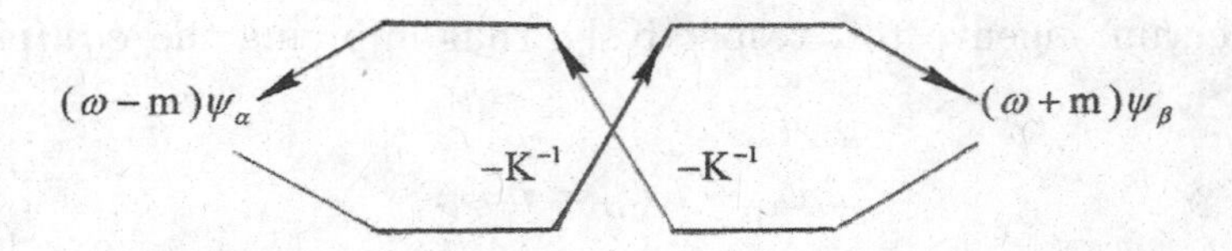

Fig. 27.2. Different display

identical to that of our "alternative model" of a boson, namely: a ***bound state*** comprised of a spin 1/2 "***fermion***" on the *left* with ***counterclockwise*** traverse and ***positive*** mass, and its conjugate "***antifermion***" on the right with ***clockwise*** traverse and ***negative*** mass. But the fermions in this case are in fact the emergent ***Dirac models*** of the ***electron and positron*** respectively!

Correspondingly we have here an interestingly ***emergent conclusion***: Dirac not only discovered ***positrons***, but he also discovered the ***Alternative model's (AM)*** version of a ***photon***! Or, we might say that the AM's model was predicted by Dirac's analysis!

So, what can we say in summary of this chapter? Well as quite likely the harbinger of things to come, perhaps we should delay our assessment of the chapter's subject until the conclusions that follow.

However, we can't let Dirac go just yet; this may also be a good place to stop for a bit and take stock of what we just went through. In the first place, we can't forget that the first order of business in this book is ***complementarity*** in terms of which what emerges from Dirac's achievement is a perfect example! What started out therein as an attempt to unite quantum mechanics and relativity achieves that purpose only if every ***fermionic*** particle has a ***complementary*** antiparticle partner! And it doesn't hurt that the analytical demonstration of that outcome's topological logic can be graphically demonstrated in a way that matches how a fermion and an antifermion are united as a ***complementary*** pair to produce a ***boson*** in our AM. And that each member of that pair can have each of two ***complementary*** spin values. Amazing! Dirac hit the jackpot!

But what's even more amazing is that all that good stuff is basically all encoded in the ***Clifford Algebra*** that Dirac was inevitably led to employ. My understanding is that Dirac did not start out with anything like all the above in mind; mainly he wanted to achieve just two things:

1. As per de Broglie and Schrödinger, he wanted to write down a partial differential equation that was first order in (the QM equivalent of) the energy and
2. To match the Klein-Gordon relativistic quantum mechanical equation with the absolute value squared of his equation.

But once he had performed #1, as we recall, it was necessary to insert the proper coefficients for the terms of his creation and he soon realized that ordinary numbers would not suffice for that purpose. Next, he tried 2×2 matrices but again with no success. So, finally, he thought why not 4×4? Ergo Clifford and the rest is history! Especially, as we see in the next chapter, that each of the Clifford matrices contain (can be written in terms of) the 2×2 Cartan/Pauli spin matrices! And finally, the way the individual Clifford matrices are deployed over the individual terms of the Dirac expression leads to their collection in the 2×2 Cartan spin matrix, the K in Eq. (27-9) above.

Magic! Much of the top-notch physics community was in awe of Dirac's accomplishment and rightly so but it was an inevitable outcome of the two requirements he set down in the first place and the empirical verification was also inevitable. ***Relativity*** is observed, so is ***spin*** and ***antiparticles*** certainly do exist. (Are you familiar with the term Positron Tomography?)

28

Statistical Mechanics

In his book, Fermi (1937) goes on to say how the original view of heat as some kind of fluid was eventually supplanted by the notion of the " — equivalence of heat and dynamical energy" which is to be sought in the kinetic interpretation, which reduces all thermal phenomena to the disordered motion of atoms and molecules. — However, what's involved is " — the mechanics of an ensemble of such an enormous number of particles (atoms or molecules) that the detailed description of the state and the motion loses importance and only average properties of large numbers of particles are to be considered."

At that point we're into the domain of ***Statistical Mechanics***, ultimately, in fact of ***Quantum Statistics***. Actually, harking back to my musings regarding the title of the chapter on dynamics, I would tend to rename the subject here as the "***Probabilistic Dynamics*** Underlying the Science of Thermodynamics (or better, ***Ergodynamics***!) as well as a number of branches of physics and Chemistry" — it's a vast subject but it started in a small way in the mid-19th century.

The pioneer was James ***Maxwell***, he of electromagnetic fame, who in 1860 derived a formula for the probabilistic distribution of the ***velocities*** of the particles (atoms, molecules, whatever; not an issue) constituting a ***dilute*** gas (implying essentially non-interacting particles) in a box with perfectly reflecting walls. According to ***Peliti*** (2011) this is an example of the simplest statistical dynamic system; a large number of ***identical***, ***indistinguishable***, ***non-interacting*** particles that, according to Maxwell's assumptions, also featured statistically ***uniform*** (and low) spatial density and a velocity density ***uniform*** in both spatial and velocity space. Besides Maxwell's pioneering work, there are a number of ways to arrive at the particle ***velocity distribution***; Boltzmann's contribution (in 1866) was essentially to re-examine Maxwell's assumptions and ***re-derive*** the distribution in a more rigorous way. In any event, here it is: the ***Maxwell-Boltzmann*** distribution!

$$f(v) = \left(\frac{m}{2\pi kT}\right)^{3/2} 4\pi v^2 e^{-\left(\frac{mv^2}{2kT}\right)}, \tag{28-1}$$

where m is particle mass, T is the absolute temperature in the box (assumed to be uniform and stable) and k is the Boltzmann constant we met before.

This is readily simplified to read

$$\vartheta(\xi) = \alpha(\varepsilon/kT)e^{-(\xi/kT)}, \tag{28-2}$$

where $\xi = mv^2/2$ is particle energy and $\alpha = \sqrt{8m/\pi kT}$.

As we see in Fig. 28.1, $\vartheta(\xi)$ peaks at $(\xi/kT) = 1$ and tails off exponentially. Also, we recognize that (ξ/kT) is proportional to the ***entropy*** as per thermodynamics of the contents of the box so that we see two things: 1. that as the box cools down its entropy ***increases*** and the ***distribution of energy*** shrinks down (which is to be expected on the basis of other analyses) and 2. that there is a most likely entropy in the sense that, at that value, the velocity distribution is a maximum.

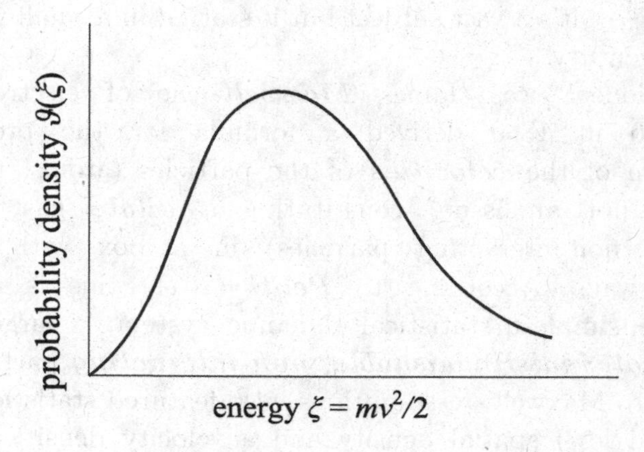

Fig. 28.1. Maxwell-Boltzmann Distribution (Not to Scale)

Actually, although it was derived 40 years later, we've already seen some Statistical Mechanics in this book; that's what ***Max Planck's equation*** was all about so we reproduce it here:

$$dE/df = Rhf/(e^{hf/kT} - 1), \tag{28-3}$$

and, as long as we're thereby into quantum mechanics, why don't we also write down, the two, also famous distributions (slightly simplified here), namely the ***Bose-Einstein (BE)*** and ***Fermi-Dirac (FD)*** distributions for particles exhibiting quantum mechanical behavior, namely

$$\mathrm{BE}(i) = 1/(e^{hf_i/kT} - 1) \tag{28-4}$$

and

$$\mathrm{FD}(i) = 1/(e^{hf_i/kT} + 1), \tag{28-5}$$

for the ith state, respectively.

A bit of commentary is appropriate here: for one thing, we note again that the exponent in Eqs. (28-3), (28-4) and (28-5) has the dimensions of (normalized) ***entropy***. Also, as far as temperature only is concerned, we see immediately that the Planck equation exhibits BE statistics, which makes sense since quantized electromagnetic waves are to be viewed as ***Bosons***. We see, also, that as temperature decreases, BE and FD statistics converge to the common value $e^{-hf_i/kT}$, eventually disappearing, but as it increases, the two diverge as

$$\mathrm{BE} \to kT/hf \quad \text{and} \quad \mathrm{FD} \to kT/(hf + 2kT). \tag{28-6}$$

And, finally, we note that the BE and FD distributions are ***complementary*** descriptors of the set of **elementary particles**, all the members of which are either bosons or fermions.

But about entropy and statistics, note that we introduced such considerations way back in Chap. 8 on Thermodynamics, beginning with Boltzmann's version of entropy, as follows (and I quote):

"For the most part, Statistical Mechanics will be covered in a chapter so named that has been delayed until almost the end of the book so as to mesh with considerations of a more comprehensive nature. Nevertheless, we can, at this point, introduce something of statistical significance that is readily seen to relate to what we have said so far about entropy, namely the ***Boltzmann*** version of that we shall designate as

$$S_B = k \ln p_B \tag{28-9}$$

where $k = R/A$ is the Boltzmann constant. R is the gas constant we met before and A is Avogadro's number, the number of molecules in a mole's worth of such so that k converts temperature to energy. According to Fermi once again, p_B is to be regarded as the probability that the ***most stable state*** of the system under consideration has been occupied."

So here we are in the chapter of reference with a statistical definition of entropy. However, as it turns out it's not the definitive definition. It's alright for the gas model that Boltzmann was concerned with but requires a bit of generalizing, something that was supplied by J. Willard ***Gibbs*** (J for Josiah) one of the most influential of 19th century American physicists[1], who took into account the possibility that a given thermodynamic system might exist in more than one, possibly a great number of, states to each of which a relative probability could be assigned. The applicable expression for the entropy is then the Gibbs formula

$$S_B = -k \sum_i p_i \ln p_i \tag{28-10}$$

where the summation is over all the possible states usually known as "***microstates***". From that point of view the Boltzmann expression, sometimes known as the "thermodynamic" viewpoint implies that there is only one state or that the probability of each of the microstates is identical.

[1] Aside from emphasizing the probabilistic aspects of Statistical Dynamics, he even coined the term!

Gibbs also introduced a so-called phase rule that, in principle, enables a determination of the number of phases to be considered in the Statistical Mechanical analysis of a given system, but it would lead us too far off track to discuss it here. Instead, here's a (very!) elementary notion of the microstates associated with the random toss of two identical coins wherein the possible microstates are

1. Both coins land heads up.
2. Each land tails up.
3. One lands heads and the other tails up.

Then, if we label the coins A and B the ***possible***, ***distinguishable*** outcomes are as follows:

A	B
H	H
T	T
H	T
T	H

all equally likely with a probability of 1/4. On the other hand if the coins are ***indistinguishable***, the probability of two heads or two tails remains the same at 1/4, but the probability of one head and one tail is now $1/4 + 1/4 = 1/2$. We could go on to other examples; many decades ago I calculated the odds involved in the toss of a pair of dice but I'm not going to do it again herein! The point is that while the physical situation can get very complicated, the basic idea is at bottom just a matter of isolating independent outcomes and tabulating recognizable groupings.

Actually, the above example is not too far from the point of view that equates Entropy to the notion of ***Information*** as a measure of the ***lack of certainty***, wherein the more possibilities that are associated with a given situation, that is to say the more ***uncertainty***, the ***greater*** the **entropy**. And, moreover, that individual probabilities add ***logarithmically*** such that if there is only one possibility the entropy disappears. In that regard, we recall the demonstration in the chapter on Thermodynamics that the

Boltzmann ***logarithmic*** expression was equivalent to the original Clausius expression for Entropy.

In that demonstration, the ***key element*** was simply the ***elementary mathematical*** fact that the ***logarithm*** of the product of two factors is the ***sum*** of the ***logarithms*** of each!

All of which brings us, as we shall see shortly, to the modern-day association of Entropy with the ***Theory of Information*** as introduced by Claude ***Shannon*** (1948). Dr. Shannon worked for the ***Bell Telephone*** Laboratories and was naturally concerned, generally speaking, with communication links that consisted of the transmission of information at one point in space, its reception at another, and its corruption by the addition of broadband "noise" in between. He was, of course concerned with how the information was, using communication jargon, encoded onto the transmitted signal, decoded from the received signal, and such system parameters as transmitted signal power and bandwidth and the relative power of the corrupting noise.

We should add that the information that Shannon cared about had nothing to do with ***semantics***, that is to say the meaning of whatever the user at either end of the link cared about; once that kind of information was "encoded" into the link it was all up to system design, be it the system telephone, radio, television or whatever!

In a way, the situation Shannon confronted was very much like that of the radar systems we discussed in Chap. 19 (if we can ignore the use of the term "encoded onto" rather than "modulated upon"). In some respects, the communication problem is more difficult because the transmitter and receiver are separated. On the other hand, a radar system ***is in fact*** concerned with "semantic" information, in the sense that it is the characteristics of target parameters that the radar is supposed to convey to the user. Actually, a number of people were quick to note that information theory has a place in the design and analysis of radar systems, a major contribution in that regard being "Probability and Information Theory with applications to Radar" by P. M. ***Woodward*** (1953), wherein the term "***ambiguity***" was defined in terms of the radar detection of targets immersed in noise.

However, the major departure that Shannon brought to the table in 1948 is the point of view that ***communication*** is inherently associated with the ***transfer of information***, something that is really quite obvious when held up to the light but as with so many subjects, something that somebody has to do sometimes! In any event, that's where Shannon and Gibbs ***coalesce***, the meeting of the minds occurring with regard to the role of probability and, again, we note it has to do with that simple ***mathematical fact***: noting that, on the one hand, the probability associated with two or more independent ***probabilities*** is known to be their product but, that on the other, we would like to have the information regarding such probabilities come wrapped in a function that must combine them ***additively***. ***Elementary but profound***!

Thus, the solution for Shannon was, of course, to precisely mirror the Gibbs expression as it was indeed for Boltzmann before them. Given n independent symbols or independent messages, with probabilities p_i, the amount of ***information*** available in the ensemble of possible messages is

$$H = \sum_{i=1}^{n} p_i \log p_i \tag{28-11}$$

where customarily the logarithm is to the base 2 and is measured in "bits" which stand for binary digits, something that many people who reads this book know all about anyway.

Let's go back for a moment to Shannon's theory of Information and the caveat that it has nothing to do with ***semantics***, that is to say with the "***meaning***" that makes life seem "meaningful", an ordered part of a pattern, etc. as discussed in the Introduction. If we feel we understand what we perceive through our senses, primarily the visual, auditory and "somato" (feeling) the raw information we receive (even as enhanced by microscopes, radar, particle accelerators, etc.), what we have done is actually to ***reduce information*** in the sense of Shannon, or, as noted by Woodward, for the case of radar signal "***ambiguity***". In some sense, we have flouted the ***second*** law of Thermodynamics! Actually, our very presence here on Earth does that!

We shall have more to say about this subject in Section VI but at this point, this is as far as I need to go in this chapter; I realize that Statistical Mechanics is an amazingly diverse and complex subject that Physicists and Astronomers have taken to the stars and beyond to the beginning of the universe, have dived with it to the depths of black holes where time stands still and filtered out in quantum mechanical tidbits. But for me, the salient feature, the key element that gives us the operative formalism is simply the elementary mathematical fact that the ***logarithm of the product of two factors is the sum of the logarithms of each!*** About as simple as you might like to see, but far reaching!

V

Some Additional Alternative Model Topics

29

"Deuteronomy" and Isospin Invariance

At this point we interject a subject of some pertinent interest to ***complementarity*** as well as to both Alternative Model (AM) and Standard Model (SM) deliberations. A model was previously shown of Pions mediating Yukawa type exchanges between nucleons to maintain deuteron stability in what may be characterized as (strong) isospin space and we reproduce it here in Fig. 29.1. This is a dynamic process, *as* postulated to maintain that stability. The figure is pretty much self-explanatory: all four ***Pions*** of the AM are involved, two for the proton and two for the neutron. At each stage of the process what was a free proton becomes a neutron and conversely what was a free neutron becomes a proton. Two fusions and two fusions take place. Also, (in what we might arbitrarily call the first stage) what was a π^- becomes a π^{0R} and what was a π^+ becomes a π^{0L}. Two fusions and two fusions take place, in each case. The process then reverses to recover the original pair of nucleons.

More generally, there is an important ***functional*** symmetry between the lower left-hand quadrant and the overlapping next diagonal quadrant of the boson matrix M shown above: the manner in which the ***vector bosons*** in the first ***quadrant act upon the electron/(anti) neutrino pair of leptons*** (in what's known as ***"weak" isospin*** space) is *identical* to the manner in which the ***Pions*** in the second quadrant act upon the ***nucleons*** (in "***strong***" ***isospin*** space).

DEUTERONOMY:
Pion exchange between Nuclei

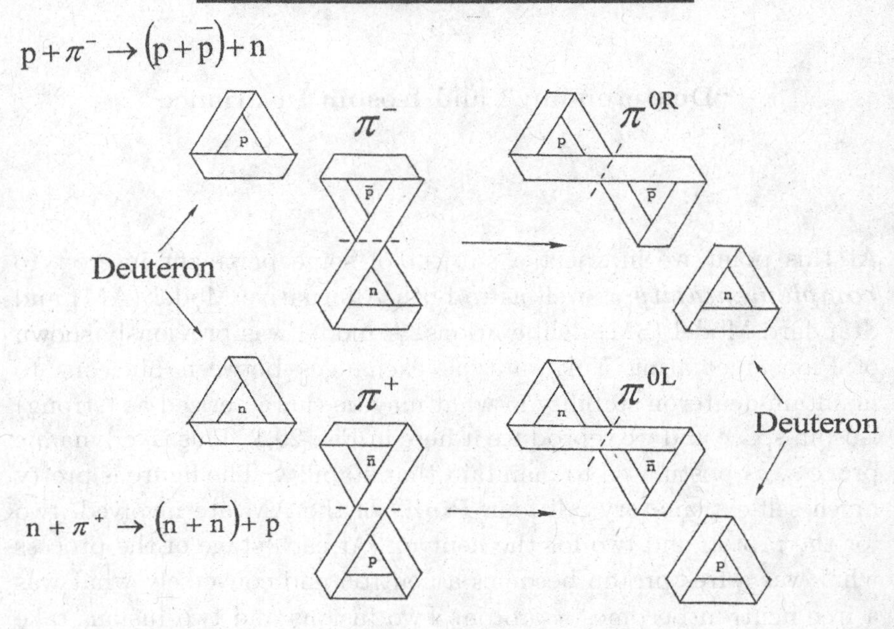

Fig. 29.1. Deuteron Stability as a Dynamic Process.

In Fig. 29.2 we show a more abstract model that fits ***both strong*** and ***weak isospin*** cases according to the identifications as listed below and the table that follows (Table 29.1) then shows the specific identifications to the two cases.

Thus in this process, the π^- splits into a free neutron and an antiproton which fuses with the original proton to make the π^{0R}, and similarly, the π^+ splits into a free proton and an antineutron which fuses with the neutron to form the π^{0L}. That is, we have the interactions that manifest what's known as ***strong isospin***:

$$\begin{aligned} p + \pi^- &\rightarrow (p + p^*) + n \\ n + \pi^+ &\rightarrow (n + n^*) + p \end{aligned} \tag{29-1}$$

such that we always have a free proton and neutron pair — basically what was presented above to illustrate the deuteron stability

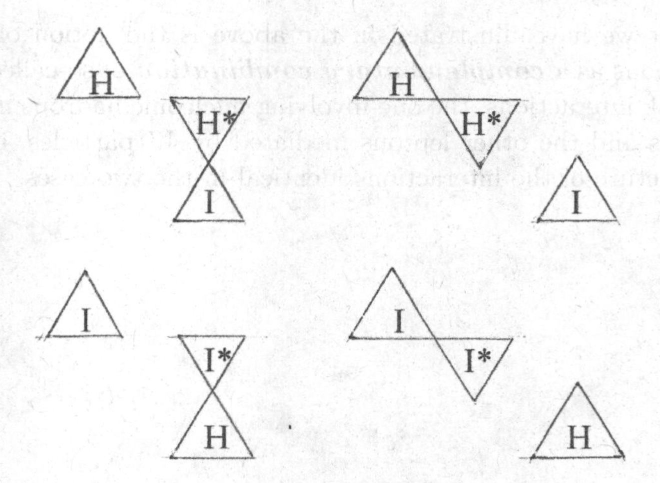

Fig. 29.2. Generalized Isospin Manipulation.

Table 29.1. Identification of Isospin

Strong		Weak
P	H	e
n	I	v^e
π^-	J	W^+
π^+	K	W^-
π^{0R}	L	Z^{0L}
π^{0L}	M	Z^{0R}

mechanism. But, now, if we make the correspondences indicated above between the ***proton*** and the ***electron***, and also between the ***neutron*** and the ***neutrino***, and so on, we can construct the ***same*** diagram for the stability of the ***electron/ neutrino pair***, with the W^+ splitting in analogy with π^- the and so on to perpetuate the electron/ antineutrino pair in accordance with the ***weak isospin*** interactions

$$\begin{aligned} e + W^+ &\to (e + e^*) + v \\ v + W^- &\to (v + v^*) + e \end{aligned} \tag{29-2}$$

What we have illustrated in the above is the notion of isospin interactions as a ***complementary combination*** of so-called strong and weak interactions, the one involving nucleonic hadrons mediated by Pions and the other leptons mediated by W particles, but with the structure of the interactions identical in the two cases.

30

Cosmological Complementarity

What's known nowadays as the Standard Cosmological Model postulates the existence of invisible "***Dark Matter***", the source of the gravitational influence that keeps the stellar constituents of our visible galaxies from being strewn about due to galactic rotation. Most literate people have probably at least seen or heard of it and are aware that it's quite unlike "ordinary matter" with which it otherwise does not interact. And that there are numerous candidates for its constituents. And perhaps also that it serves a vital cosmological function: According to current expertise, dark matter seems to be distributed in a kind of grid-like fashion throughout space and many of our galaxies appear at the intersection of the grid lines. So, although we can't see it, dark matter is a vital constituent of the cosmos we know about.

You may also be aware of the current (as of this writing) search for evidence of what's known as ***Supersymmetry***, that is, the expansion of Standard Model (SM) elementary particle taxonomy to include the existence of a bosonic partner for each fermion and, vice versa, a fermion for each boson. There seems to be an element of what might be described as desperation to the search! According to Lykken and ***Spiropulu*** (2014), "Supersymmetry is an amazingly beautiful solution to the deep troubles that have been nagging at physicists for more than four decades. It provides answers to a series of important "why" questions: Why do particles have the masses they do? Why do forces have the strengths they do? Why does the Universe look the way it does? — most of the world's particle physicists believe that supersymmetry ***must*** be true, the theory is that compelling."

Thus its' apparent absence to date seems to pose a threat to some of the SM's basic tenets, including a relationship to particle mass and, more to the point of this Chapter, the source of Dark Matter.

I would like to offer the particle physics a ray of hope; as far as mass is concerned, we talked about it in a previous chapter in terms of our "***indigenous parallel***" to the ***Higgs*** potential. And beyond that, here's what we might call an "***indigenous* model**" for ***Dark Matter***, admittedly speculative, but emerging from our alternative model in a logical, almost compelling way: Always lurking at the perimeters of our earlier deliberations but never really contributing to them is that enigmatic ***fourth*** elementary fermion, the one with a twist of +3 and a charge of +2. Clearly, this FMS as well as its taxonomical combinations, the triad D, C and B have little to do with the alternative model's taxonomy or interactions. Nevertheless, they ***must*** appear in our modeling by virtue of symmetry and so they do, at each level of fusion, a kind of ***super (anti)symmetric*** manifestation of ***complementarity***.

That being the case, the question naturally arises as to whether those super (anti)symmetric particles might actually have as much ***relation to reality*** as do our elementary fermions A, B, and C and their taxonomical combinations. So, let's assume that ***our*** dual particles actually do ***exist*** but that, in the manner of Dark Matter, we can't "see" them simply ***because*** they do not interact, at least in any kind of electromagnetic way, with normal matter. Though they are therefore "dark", as Solitons, "in and of" spacetime, they should be endowed with mass, possibly very heavy mass, as inferred from the apparent gravitational influence that led to their putative existence in the first place, in which case the ***nature*** of that putative endowment is the issue. As long as we're in a speculative mode we could invoke the notion, as discussed in a previous chapter, the ***triplication*** of time as part of the explanation for additional generations of particles. Although, conceptually, such triplication should exist at every point of space, there appears to be no prior reason for a "*dark*" particle's ***ordinary*** time to be the same as it is for ***our*** ordinary taxonomical world. That is, it could instead be one of the other orthogonal time coordinates in the triplication, perhaps the one associated with our

heaviest generation in which case, finding a "dark" particle might be very difficult.

But here's another idea that will make the above seem a bit conservative: to begin with, perhaps you've also seen the recent article in Scientific American about the way in which the matter we know and love, that is to say "ordinary" matter — let's call it "***our*** matter" — is distributed in space. The article discusses how simulations of the universe, using all "known" aspects of how it evolves, end up with dark matter spreading across space in a network of filaments — "a cosmic web" — at the intersections of which galaxies, especially dwarf galaxies, of "***our*** matter", at least those we see, ***tend*** to form (there are complications which is what the article is about, but that are of no immediate interest to us).

Actually the notion of the cosmic web of filaments is not new but in any event, here's my idea: I'm going to ***assume*** that the universe we encounter is, in reality a ***composite*** of ***two***, ***complementary***, "Component Universes": one involving "***our*** matter" and one "***their*** matter", with "their" translating as "***dark***" for us. That's it; two component universes; *Ours* and *theirs*; *Light* and *dark*; *Yin* and *Yang*! As per the thesis of this book, ***Ours*** is built from the elementary fermions A, B and C and their conjugates, as well as the taxonomy that results from their combination. What I postulate here is that ***theirs*** is similarly built from the fermions D, C and B and the associated combinations. The two components influence each other gravitationally — obviously dark matter influences where the light matter congregates as per the above.

However, although, as we observe it right now, the two universal components coexist, I further ***postulate*** that they may not always have done so or will do so; they have different ***histories*** — Cosmic "life cycles" as they say. In fact, one or the other may, in some sense, be in an ascendant phase and the other decaying. For example, let's suppose that the dark component has been around much ***longer*** than our light one. The implication of course is that the ***former*** was the host for the emergence of the ***latter***, a situation that would appear to answer the question as to what came ***before*** the Big Bang! Furthermore, it would render ***moot*** the either-or relationship between the steady state and big bang scenarios for our universe; the answer is ***both***! But of course, it is the dark component that presently dominates and determines the underlying ***structure*** of *our* universe and, if our hypothesis is correct, did so for a long time, even before our component existed. And, not unlike electricity and magnetism, constitutes one, complete entity. And as far as we are concerned, the dark component is indispensable for the continued existence of our universe, at least in the form that permits our existence!

At any rate, "apparently", (meaning the Standard Cosmological Model's (SCM) best current estimate) about 14 billion years ago, what we've been calling herein ***our*** universal component began. Most of us are reasonably familiar with the SCM's Cosmic Life Cycle story; a monstrous "inflationary" phase, followed by "ordinary" expansion with the emergence of particles, background radiation, stars, galaxies, etc., — there's little to be gained by recounting it at this point. However, there is one more point that ought to be stressed: the implication of what we've been saying here is that not only did our component begin at a particular ***epoch***, it must have begun at a particular ***location*** in the ***already existing dark*** component universe. Of course, there's no way we can specify what that location is; all we can say, in other words, is that our component universe began at a ***particular spacetime*** point in an ***already existing*** spacetime. That may be a fine distinction, but it differs from the usual point of view that, in essence, time and space grew, along with the expansion of the universe as we know it; after all, we're not the sole arbiter of that matter. And a word of caution: location of the point of emergence in the ***Dark*** universal component does ***not*** mean that we can discern a similar location in ***our*** component universe; being part of it forbids that!

And just so you don't get the wrong idea, in some sense, what I've been describing is, of course, not really new in its entirety; the notion of continual emergence of new universes from preexisting old ones has been around for some time, having originated with a number of cosmologists including ***Vilenkin*** and Andrei ***Linde***, also one of the pioneers of Inflation theory. Nevertheless, it is hereby asserted that this chapter has introduced a new wrinkle into the grand and glorious fabric of cosmological speculation. It would of course be nice to have a convincing sequence of mathematical argumentation to lend credence to the wrinkle but, again, maybe next time.

A couple of final remarks: first, both of the above universal components are dynamic entities and, although at this point we don't understand it, the interplay between them includes, one would expect, some kind of oscillatory process that we may or may not talk about! And last but not least; we note that our two universes arising, inevitably, out of our initial formalism can be viewed as a sort of ***cosmic manifestation of the Principle of Complementarity.***

31

Gurule Loops

I really hadn't intended to write another substantive chapter before the concluding section but here it is — my history repeating itself! The bit of history involved here is that when my Book of Reference (BoR) was almost complete, I realized that a simple 2×2 matrix can encapsulate the essence (as symbolized by the Principle of Complementarity) of ***both*** the elementary particles of physics and DNA, that marvelous, elementary molecule that enables ***both*** the inherent multiplicity of life forms on earth and their evolution with time. That realization was quite mind Boggling, a condition so dire, as to launch my quest into the ***range*** of that Principle's applicability — that is to say, where and how fundamentally it applies; ergo the current book on the Principle of Complementarity itself.

This time it's not quite the same situation but similar enough for me to tell you the story of this chapter. In the first place, there was always the nagging possibility that another such complementary pairing might exist somewhere in the immense "lacuna" between our original pair. Again, however, time was wasting; the book seemed to have quite enough material to publish, and no obvious pairing was presenting itself.

And then, just as in the movies, (or maybe in a detective story!) here comes the gimmick[1], the little unforeseen item that launches the story. This time it was a phone call, right out of the blue that comes along suggesting that the reluctant pairing might occur between

[1]There's an applicable phrase here: Maybe "Deux ex Machina"?

atomic orbitals and Gurule Loops! (Gurule Loops?) And then (and if) I were to act upon the suggestion, the overall result would be to increase the aforesaid "range" to the complete span all the way from the elementary particles to "real life" humans! So; the phone call, it appears, might turn out to be most opportune — Amazing!

Of course, atomic orbitals have been recognized for quite some time and are well-understood by the relevant scientific community; in general terms, an AO is mainly just the number of electrons in the specified atomic orbit for a given element. On the other hand, Gurule Loops are an entirely different matter for most readers as they were for me when early last year (meaning 2017) I received that mysterious phone call described above. It was from one Stephen Taylor, who, although his background is primarily in music, is involved, along with Jesus Trevino (whose given name is pronounced Hay-soos), a retired movie director and technician, in the production and dissemination of a short, special-subject movie all about those loops and their inventor/discoverer, one Benjamin Gurule, primarily an artist and sculptor but apparently possessed of that curiosity that begets novel fields for study.

Although Gurule, himself, is no longer with us — a tragic loss — he was able to give a number of demonstrations of his loops and their potential, which brings us back to the story of the phone call. Which, as you might expect, requires some amplification and for that we bring back Taylor and Trevino, both ardent scholars of the Gurule Loops phenomenology. In fact, they met at Gurule's memorial service where they agreed to honor his request that his work be kept alive and provided with a wider audience. The result of their discussions of the matter was that the two decided to make a film dedicated to that explicit purpose and the phone call was associated with the business of the film. But, the question remains "Why did Taylor call me? What could I possibly have to do with "the business of the film?" " (To say nothing about Gurule Loops, explicitly, anyway).

Well, here the plot thickens a bit and at this point we must introduce another key player, namely Professor of Mathematics, Louis Kauffman, UIC, well-known globally for his expertise on the theory of knots (among many other abstruse subject matters). But Kauffman has another side to that expertise; he applies it to the discovery and encouragement of not-well-known but worthwhile individuals who find it difficult to have their contributions to the advancement of science recognized, an activity I find most laudable, especially since I like to count myself as one of the beneficiaries of his activities in this area, something for which I am eternally grateful.

But to close the loop (no pun intended), it turns out that Lou was also quite aware of the Gurule loops, which, most likely caught his attention as also "worthwhile". So, when Taylor visited him in Chicago, the discussion was primarily about the loops. However, at one point it turned to the subject of knots because — and here my understanding gets rather hazy, from what I can gather, Taylor related his impression that one can discern on the surface of a sphere (see below), well-adorned with Gurule loops, something that looks like one or more of my flattened Moebius strips!

What, specifically, caught Steve's attention was what is called in the Gurule world, a "***convergence***" meaning, in this case, the combination of several "loops" to form the sides of an apparent MS. So, although convergences were not planned to be the main subject of this chapter, let's let Stephen finish his story: As he recalls, at this point in his visit, Kauffman says "wait here", disappears into his inner sanctum and comes back with a copy of my original book, my BoR! But, to (finally!) have this story come to a satisfactory ending, Kauffman also advises Taylor to get in touch with me, ergo the phone call, a subsequent visit by Taylor and Trevino (T and T) to my home and, eventually, my promise to take a close enough look at the subject of Gurule Loops to write something cogent about it. Fortunately, T and T at some point also provided me with a DVD showing one of Gurule's demonstrations featuring his loops on a spherical last and so, that's what I've used as a reference for what follows.

Of course Stephen's main focus at that point was still convergences, so, with an eye towards truly understanding the phenomenology, Stephen has since, wisely resorted to actual, physical reconstruction of Gurule loops on a sphere and, as a result, he recently sent me a box containing some specific reconstructions of them which are indeed most edifying.

Now, let me explain the situation Stephen faced in such reconstruction: in the first place, the task was to put together a set of identical loops, in a certain arrangement, on the surface of a sphere that in reality will ***not actually exist*** but whose existence will be ***implied*** by the ***actual*** existence of ***that*** set of loops that ***would*** actually adorn the surface of the sphere, were it to actually exist!

That ought to clear things up, right? Furthermore, the particular arrangement of the loops was dictated by the need to exhibit a certain characteristic, which as per Stephen's original concern was the notion of ***"convergence"***. So! Stephen solved the problem and his solution was an exercise of genius!

To begin with, I surmise, he made a trip to his local supermarket to purchase a set of those very slender plastic "straws" customarily used to slurp up liquid goodies. Those straws are about ***22*** inches long so that when bent into a circular shape, attached, head-to-toe and properly emplaced they end up enclosing the ***implied existence*** of a ***sphere*** about ***7*** inches in diameter.

The "sphere" Stephen then sent me was apparently constructed out of 6 such straws of various colors but one is best identified as the ***equatorial circumference*** of the implied sphere so that only 5 straws are left to carry out the business of illustrating convergence which requires that the loops be used in pairs: Ergo, two such pairs to begin with and one maverick loner. Nevertheless, using various combinations of loops, we end up with exactly what is required and more. Of course, this requires, not just double, but triple and for a few loops quadruple duty, which (I think) is possible only on a spherical surface.

In any event, what do we see when we examine Stephen's creation? Well, although all loops must go completely around the imaginary sphere (that's why they're called "loops"!) let's just consider the "Northern hemisphere" starting right above the Equatorial Circumference of the sphere where, because we are looking for triangles, the two members of each of the pairs I mentioned must cross each other (which they do) ***twice*** per rotation, once above and once below the Equator, the two members of each particular pair being separated in vertical orientation at the crossing by about ***45 or 50*** degrees. So, each member is oriented half that from the vertical (the N–S direction).

As a result, if we circle around the equatorial circumference, but constrained to the Northern hemisphere, we encounter 5 "vertical" double triangles (Fig. 31.1) whose "bottom" is provided by the circumference itself and whose top is provided by one of the other loops. We note that the equatorial straw actually winds in and out of the sides of each convergence so that if we're navigating from West to East the Equator goes "under and over", under and over, etc. These "double triangles" are exactly what is in our Book of Reference which: are known as the result of the "First order fusion" of a fermion and an antifermion and, in this particular case, the fermion is an electron, the antifermion is a positron and the combination is the model for a photon as employed in Chap. 23 to model instantaneous, "long-range phenomena".

But to return to our original subject matter, we see that the two members of each paired loop are relatively closely spaced at each double triangle so that any two such must be separated by a sizeable space with a certain formation. What kind of formation? Well, it turns out to be ***Pentagonal***! (Yes, pentagonal just like that famous building in Washington, DC!) Two sides of which are provided by the "adjacent" double triangles, the bottom by the equator and the top by another loop. So, as we course around the equator, we encounter 5 double triangles and 5 pentagons, first one and then the other — triangle, pentagon, triangle pentagon, etc. And we should note that the two sides of the double triangle widen up into a ***pentagon*** as we follow it in a ***southerly*** direction with double triangles also oriented in a southerly direction on either side of it!

So, since the Southern hemisphere is essentially just like the above, it would appear that that's what Stephen Taylor's reconstruction is made of; just ***those configurations:*** a combination of 10 real, ***double triangles*** made of plastic "straws" and 10 ***imaginary Pentagons***, made of the "nothing" between the circumference and the adjacent configurations. And, that's it; that's what we can expect to see, right?

Well, not quite. That's it except for two heretofore ignored additional terms, that is! I call them the ***Polar pentagon*** and its partner the ***Subpolar Pentagon***! Heretofore, we have ignored the fact that our paired loops continue up and over the polar pole. And down and under the subpolar pole. So those two pairs enclose and define the two polar Pentagons, except for one open length and for that we see that our maverick extra straw we mentioned above. In other words, the top ends of the double triangles form the sides of the Polar Pentagon! And similarly for the bottom ends of the double triangles for the subpolar Pentagon.

Well now that really is "***IT***", no more surprises. So, let's take stock (so to speak): we (actually Stanley Taylor) started with half a dozen slender, plastic straws, each of which has ended up as a loop around an imaginary sphere whose "surface" appears to be covered by a pattern of "double triangles", regularly interspersed by spaces with a pentagonal shape. In fact, if we proceed north starting with a double triangle we get two pentagons (One being the polar pentagon) followed by a southerly double triangle and finally two southern pentagons whereupon we're back to the beginning. Which is another way to summarize the organization of the whole Gurule sphere!

But to return to the ***significance*** of the sphere; as noted above, the double triangles may be regarded as models of the first-order mating of a Fermion and its antifermion partner and thus something to be recognized as "***real***" in some sense. On the other hand, the pentagons appear to exist only because the spaces between two adjacent triangles actually have the form of a pentagon; take way the triangles and the pentagons disappear!

However, the triangles cannot exist, sans pentagonal spaces between them, either! In other words, the ***triangles*** and the pentagons must ***coexist;*** they are ***Complementary*** phenomena just like the celebrated ***Yin and Yang*** symbols on the cover of this monograph and thus may be regarded as valid subjects with which to end this, unfortunately, too brief discussion of "***Complementarity***" as it relates to Gurule loops (or vice versa!).

But wait! Let's take a larger look at Stephen Taylor's Gurule "Sphere". And we certainly *recognize* it as a sphere or, more precisely, a spherical ***shell***, but in any event, now we see an entire ***Equatorial band*** of ***Complementary*** phenomena recognized as such because of the two ***Complementary*** Polar ***Pentagons*** that hover above and below it! More ***complementarity***!

Yes, more, but not all of it! Don't forget that, as a spherical ***shell***, Stephen's creation has an ***inside*** as well as an ***outside*** and those two entities are ***also*** in a ***complementary*** relationship, albeit a higher dimensional one!

Well, this time, that really is "***it***" and (I believe) we have exhausted all the ***complementarity*** Steve's creation has to offer! So let's change the subject a bit: Does all this talk of Equators and Polar Regions remind you of anything? I thought so and I have a model of Planet Earth in our living room (of the humble abode Charlene and I have lived in for 45 years or so). I see great land masses of that globe surrounded by oceanic regions that, I understand, somehow ***evolved*** over countless millennia whereas our model evolved in Stephen Taylor's mind and hands. Is that something to think about? Well, maybe, but that's probably as far as we ought to carry that kind of speculation herein.

So, before we end the discussion, let's pause a bit and consider an implication or two of the preceding. You know, a few weeks ago, I showed Stephen Taylor's creation to some family members and a friend of the family — let's call her Bonnie — commented that she didn't understand all the talk about triangles; what she saw were pentagons! So, suppose you had an actual spherical shell (Never mind the material!) and you cut pentagonal windows out of it just like they appear on our Gurule sphere; would someone not be justified in claiming they saw some interestingly-shaped triangular constructions and wondered what they were made of?

By the way, Stephen actually constructed a set of seven Gurule spheres as we see in the figure below (Fig. 31.1). Each contains a black balloon blown up to make visualization better and it appears that Q6, the one with six encircling elements is indeed the best combination

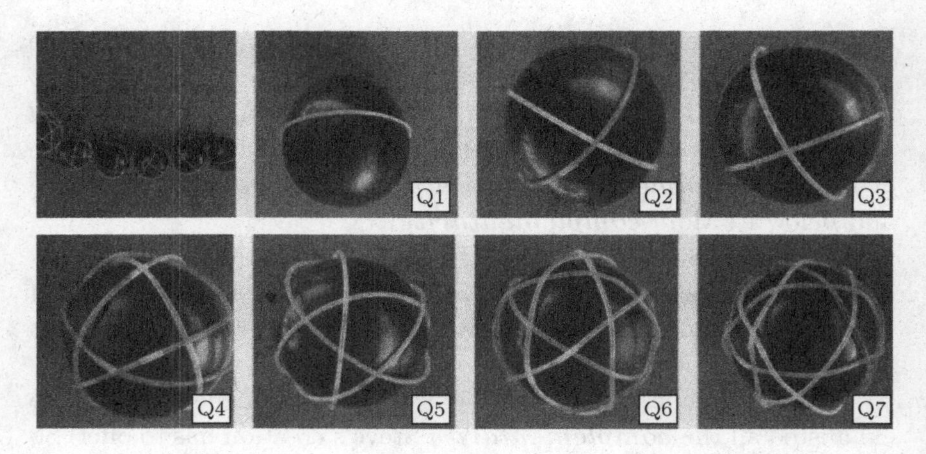

Fig. 31.1. Seven Gurule Spherical Candidates for Analysis

of regularity and diversity for purposes of discussion. In fact, it is so symmetrical that the selection of a particular plastic straw to serve as the Equatorial Circumference is completely arbitrary; all the geometrical relationships discussed above would emerge if any other straw were elected.

Think about that as we segue to the next chapter and some concluding remarks for this book. For one thing, it occurs to me that, in this chapter, we have, "willy-nilly", been operating in the bailiwick of what has heretofore been entitled *The Meaning of "Is"* which, you may recall, we discussed a bit early on in the book. I'm sure that if you think about it you'll agree. So, we now proceed to the summation and interpretive phase of this document but with something else to add to the discussion that we didn't have before we wrote this chapter! Perhaps Benjamin Gurule would have been highly gratified!

VI
Summary and Conclusions

Nothing exists from whose nature some effect does not follow
Benedict (Baruch) Spinoza

32
Recapitulation

If you read the Introduction, you may recall a quotation attributed to Professor John D. Barrow (2007) regarding his discussion of the quest for "The theory of everything" and repeated here as follows: "... If we are to arrive at a full understanding of complex systems, especially those that result from the haphazard workings of natural selection, then we shall need more than current candidates for the title "Theory of Everything" have to offer. We need to discover if there are general ***principles*** that govern the development of complexity in general which can be applied to a variety of different situations without becoming embroiled in their peculiarities." (My emphasis).

To which I asserted that "I think the ***Principle of Complementarity*** qualifies as such for a wide variety of situations —." Actually, from an epistemological point of view, the strategy of writing the book was, at least to begin with, basically "empirical" in the sense that I wanted to demonstrate a ***complementary*** underpinning for as many individual subjects of the physical world as possible, thus providing a basis for a generalizing ***assertion***, in the absence of a "Theorem" or logical "proof" to that effect, that Complementarity is indeed a ***universal principle.***

Well, I haven't given up on the possibility of such an unassailable formality as a "proof" (see below) but it seems to me that we have come a long way in the preceding chapters toward providing such a basis, having examined for corroboration (successfully, by the way) at least a couple dozen subjects, give or take. Interestingly enough, however, it occurs to me that the two most important subjects are those I borrowed from the previous book, one on the nature of DNA and the way it replicates, and the other on my Alternative Model (AM) of the elementary particles. As I indicated in the Preface (or was it the introduction?) I thought it was important to show how it was that I got started on my mission to give ***Complementarity*** its rightful position in the Pantheon of Principles, so the first two chapters of this book after the introduction are mainly a reprise of the treatment in that first book.

Chapter 4 was then a comparison of the two topics with the primary result being what I like to think of as the ***signature*** of ***Complementarity***, at least that kind of complementarity, a spare little 2×2 matrix I call m, with a haunting resemblance to the Yin-Yang symbol. However Chap. 5 was indeed something new: it treated my little signature matrix to a tiny bit of algebra that in a few lines illustrates how something like m can replicate itself ad infinitum! There is a short discussion in that chapter about the limiting requirements for an m-like replication process to proceed but nothing about its larger implications so I shall return to that subject momentarily (See below). Anyway, the point is that by itself (I'll talk about replication later), matrix m may be viewed as an ideal against which to compare additional topics. And, of course, such topics abound.

Starting with Chap. 6 we then proceed to cover, to one extent or another, Dynamics, Thermodynamics, Electrodynamics (mainly Maxwell's Equations thereof), Special and General Relativity, Differential Geometry as it contributes to the AM[1], Noether's Theorem and Gauge Theory, Quantum Theory (in several parts), Pauli's spin and Dirac's Antiparticles. I should hope everyone realizes that there is no way in the world I could even approach anything like complete coverage of the world of Physics in this one book. And if anyone's favorite topic is missing it is not an intentional slight; mainly it bespeaks a paucity of familiarity with the subject or a lack of quick reference material!

---

[1]With the unearthing of an unexpected subsidiary conclusion that is bound to be controversial!

At any rate, I also like to think that the coverage is enough to give a bit of a boost to Niels Bohr's insightful principle so let's leave it at that for now, and let me segue over to that subject I inserted into the title of the book, namely, the enigmatic ***Meaning of*** "***Is***", phraseology that, I hereby stipulate, was not (at least primarily) added to the title in order to stimulate interest in the book (although it can't hurt if it does!). On the contrary, its inclusion is legitimate but to make sure I looked up *The Meaning of "Is"* in the dictionary and I got the grammatical meaning, to wit; "Third person singular" and Present Indicative of "be" both of which sound familiar but it's been many decades since I was exposed to that sort of thing. So I looked up "be" and Lo and Behold, there are over four inches of meanings! The one I like the best is "***to exist***" and, for me, it has to do with how we view Space and Time, and in that regard, I believe the ideas of the philosopher Benedict (Baruch) ***Spinoza*** extends thereto.

As I talked about in the previous book and I'm sure you know, Spinoza was a very influential 17th century philosopher and his concept of "***Monism***" which is that there is "only one, universal, absolutely infinite fundamental substance" (Nadler 2020) intrigued a lot of big thinkers including Albert ***Einstein*** who, as related by John ***Wheeler*** (Avrin 2015, p. 196) believed that, someday, Space and Time will be recognized as the ultimate reality. Also in the previous book is a whole page devoted to the 19th century ideas of William Kingdon ***Clifford*** (renowned for Clifford Algebra) who, impressed by Riemannian geometry, envisioned ***matter*** as disturbances in space — "little hillocks" — moving in what we would recognize nowadays as a ***solitonic*** manner to occupy successive portions of that space (Clifford 1870).

So, in summary, here's how I interpret *The Meaning of "Is"* at least with regard to the subject matter of this book: In consonance with the visions of Spinoza, Clifford and Einstein, I assert that

1. There "***is***" only ***one elementary substance*** and it is ***Space-time***, a concept created by the ***complementary*** unification of Space and Time by Einstein's Special Theory of Relativity and Minkowski's four-dimensional ***geometric*** assessment thereto.

2. Further, otherwise unaltered ***Spacetime*** and the ***Matter*** by which it is inhabited are ***complementarily*** unified by the concept of the localized ***Solitonic*** distortion of spacetime energy to create the elementary particles of the Alternative Model via a process that equates General Relativistic Spacetime ***curvature*** with local ***Stress-Energy*** as discussed in Chap. 11.
3. In terms of an ontological ***point of view*** of the elementary particles, we have thus arrived at a stopping point: ***Space, Time and Matter*** are united and particulate ***reductionism*** can proceed no further.

That certainly doesn't mean, however, that we have the answers to all the questions that have been or can be posed about the relationships of space, time and matter, nor do we even know if we can yet "frame" all such questions; we are nowhere near "Ultimate reality". But here's what I believe we can say.

33

Linking Complementarity and The Meaning of "Is"

What follows takes a while to expound so pleased bear with me: With Spinoza and Clifford in mind I ask rhetorically, what can we say about *The Meaning of "Is"* and the way in which it relates to ***Complementarity***? In that regard, I would like to present the following "semi-axiomatic" chain of statements:

1. Let us ***assume*** that the word "is" implies the existence of ***something*** rather than ***nothing***, words whose meaning we purport to understand. (Axiom 1).
2. ***Assume*** further that Spinoza was right: there is only one "***something***", a, universal, absolutely infinite, fundamental ***substance***. (Axiom 2).
3. Now identify that "substance" with the Einstein/Minkowski ***Spacetime***.
4. Further, identify Clifford's local disturbance of that spacetime with the solitonic "***particles***" of the Alternative Model.
5. Note that, as ***Solitons***, these entities display ***both*** particle-like and wave-like characteristics and thus manifest their own "***Complementarity***" as per Bohr.
6. Note, also that the existence of the ***question*** as to *The Meaning of "Is"* implies the concomitant existence of a live, sentient ***questioner***[1], which, at least in our world, implies the associated existence of cells, composed of molecules, composed, in turn of atoms, and, ultimately, ***elementary particles***.

[1]Please don't accuse me of sophistry.

7. Which in the present context, implies the ***Solitonic*** particles of ***the Alternative Model*** for which ***Complementarity*** was shown to be a fundamental ***attribute.***
8. The implication is thus that the ***Principle of Complementary*** is a necessary, unavoidable partner of *The Meaning of "Is"* where the latter is defined as per Axiom #1 as *the existence of* ***something rather than nothing***: the two are inextricably linked and at a ***cosmological*** level!

That's it: My "Point of View" is that *The Meaning of "Is"* and ***Complementarity*** are ***Cosmological partners*** in something like ***Ultimate Reality*** or at least a version of it for which the existence of ***something rather than nothing*** is, I believe, as fundamental a definition as one can imagine. I realize, of course, that my ***Alternative Model*** of the elementary particles is also a **necessary** ingredient for the validity of the above sequence as is the necessity for the existence of ***sentient beings*** to experience the Cosmos and to ask the necessary *questions.* I guess we can qualify for that.

But, notice: the above sequence puts us squarely in the picture as ***necessary*** ingredients of discussions of a ***cosmological import***. Who was it that said, "I think therefore I am"? I seem to recall Descartes (Hatfield 2018). Anyway, perhaps we may be excused for our brazen extension of his proclamation to read "I ***think*** therefore the ***World is***." Which would seem to suggest that those who insist on ***Mankind*** being the ***goal***, the "be all and end all" of the ***creation*** of the Universe, a point of view I talked about in my previous book, are on the right track.

On the other hand, again, as I also mentioned in the previous book, those terminally ***optimistic*** individuals find themselves squarely up ***against*** the ***notion*** that our ***Universe*** is ***not*** necessarily ***unique***: that, in fact, countless ***others*** may "***exist***" with countless other sets of ***parameters*** such that a universe such as ours, featuring creatures such as us, at least in terms of being sufficiently "***sentient***" to ask similar ***questions***, is ***statistically*** bound to emerge — no big deal!

But what happened to the subject of ***Gurule Loops***? Well, it also has something important to contribute to *The Meaning of "Is"* but the way it does so is not as easily described. Perhaps the most straightforward way is to begin with our family friend, Bonnie and her ***recognition*** that what was ***apparent*** to her about Stephen's creation was the existence of ***Pentagons*** rather than Triangles!

Now, the fact that those inter-triangular areas do indeed have a pentagonal shape is not the issue here. What ***is*** the issue is Bonnie's ***recognition*** of it, a ***mental process***! And what is important to this Final Chapter is the ***need*** for that process.

But hold on you say? How about our carefully structured algorithm leading to the role of the Principle of ***Complementarity*** as a necessary, unavoidable partner of *The Meaning of "Is"*? That requires an observer, right? Well, perhaps Bonnie will do well in that regard!

And are our double ***triangles*** and the ***Pentangular*** formations between them not avowed ***Complementary partners***? I feel we put that requirement to rest in the Gurule Chapter. As far as *The Meaning of "Is"* is concerned, Benjamin Gurule must be smiling in his grave; he knew it all along!

So what do you think? Or do you? Here's what I think (or at least tend to think!): We are indeed important but, initially, as species, we have been so for most of our existence, but only to ourselves and no more so than the other inhabitants thereof. And our Universe, our solar system, our planet were all similarly important to us. But now, and it has happened very, very quickly in the cosmic time scale of things, we are in trouble, dire trouble. Our ability to ask cosmic questions, you might say our scientific and technological capabilities have ***expanded*** enormously which, you would think, is a good thing. However, as it is with such things, there are side effects, in our case also enormous side effects. We have become important to our planet — the entire planet. And to all the inhabitants thereof, including ourselves, frighteningly important and the future, including the civilization that has spawned all that scientific capability doesn't look too good right now. I'm sure you know what I'm talking about and I'm sorry to leave you on such a dismal note.

But in any event, in terms of the purpose of this book, perhaps we made a valid case for the Principle of Complementarity as a Universally Applicable Criterion for Elementarity. Let me know what you think.

javrin@aol.com.

Appendix: The Basics of Beta Decay

(Including the transformation of the Neutron and how Pions operate) To begin with, we are specifically concerned with the *entire **vector** of four* basic spin1/2 *fermions* $V = (A, B, C, D)^T$ plus its conjugate vector of *antifermions*, $V^* = (A^*, B^*, C^*, D^*)^T$ and their direct product

$$M = V \otimes V^{*T} \tag{A1}$$

the result being the (self-adjoint) matrix of sixteen two-element composites shown in Eq. A2 where, in analogy with quantum mechanics, we note that M can be viewed as an ***operator***

$$M = \begin{bmatrix} [AD^*] & BD^* & CD^* & DD^* \\ AC^* & BC^* & CC^* & DC^* \\ AB^* & BB^* & CB^* & DB^* \\ AA^* & BA^* & CA^* & [DA^*] \end{bmatrix} \tag{A2}$$

Where:
A is the Electron (twist $= -3$, Charge $= -1$)
B is the Neutron (twist $= -1$, Charge $= 0$)
C is the Proton, (twist $= 1$, Charge$= 1$)
And D is the equivalent of the Electron for what we expect will one day be recognized as the so-called "Dark Matter" (Twist $= 3$, Charge $= 2$).

In 1948 Enrico Fermi and C.N. Yang wrote a short paper (Fermi and Yang 1949) suggesting that that the ***Pions*** are bound states of a nucleon and an antinucleon — and ***there they are***; the ***four terms grouped in the center of M***!

We'll talk about them in a minute but first, let's take care of the decay of the Neutron into a Proton, an Electron and an

Antineutrino which is a classic example of a "***weak process***"; it's quite complicated (and anyway requires **Pions** to make it happen!). So, for the neutron, we can illustrate the process in terms of the ***unique*** (As far as I know!) diagram shown in Fig. A.1 where we begin with Twist = −1 (Minus sign meaning it's to the left),

1. It is then folded so that, while the Twist still = −1, the Charge becomes −1 as well.
2. The folded arrangement is then rotated, simply for convenience of viewing.
3. An available (!) charged Pion (CB*) is then mated with the folded arrangement as shown, that is, one of the two up Quirks of the Pion is mated with the available up Antiquirk of the folded Neutron so that we then have total Twist = 0, Charge= +1.
4. Once the three cuts shown are made we are left with a Proton, an Electron and an Antineutrino as expected (per experimental results) plus a ***different***, uncharged pion (CC*). The net result has total Twist = −1 and charge = 0 which is ***what we began with!***

Finally, there's something else we need to talk about before we end this Appendix, something ***unanticipated*** but ***most important*** (Believe me!).

So, let's go back to, the ***Pions*** for a minute; we only used two of them, CC* and CB*, but they are ***complementary*** and must be so considered. However, what we ***really*** want to talk about applies to the ***complete set of four***.

So, let's suppose we use the above ***numerical values for the twist of B*** and for ***both*** the ***twist and the charge of C*** (B has no charge anyway!) We then obtain the somewhat ***unanticipated*** (!) ***result*** for our set ***of four Pions***, as shown

$$\begin{pmatrix} \mathrm{BC}^* & \mathrm{CC}^* \\ \mathrm{BB}^* & \mathrm{CB}^* \end{pmatrix} = \begin{pmatrix} 1 & -1 \\ -1 & 1 \end{pmatrix} \tag{A3}$$

That is, when considered as a ***2 × 2 matrix***, it is ***transformed into our*** **ubiquitous Matrix** ***m,*** the matrix that, among other

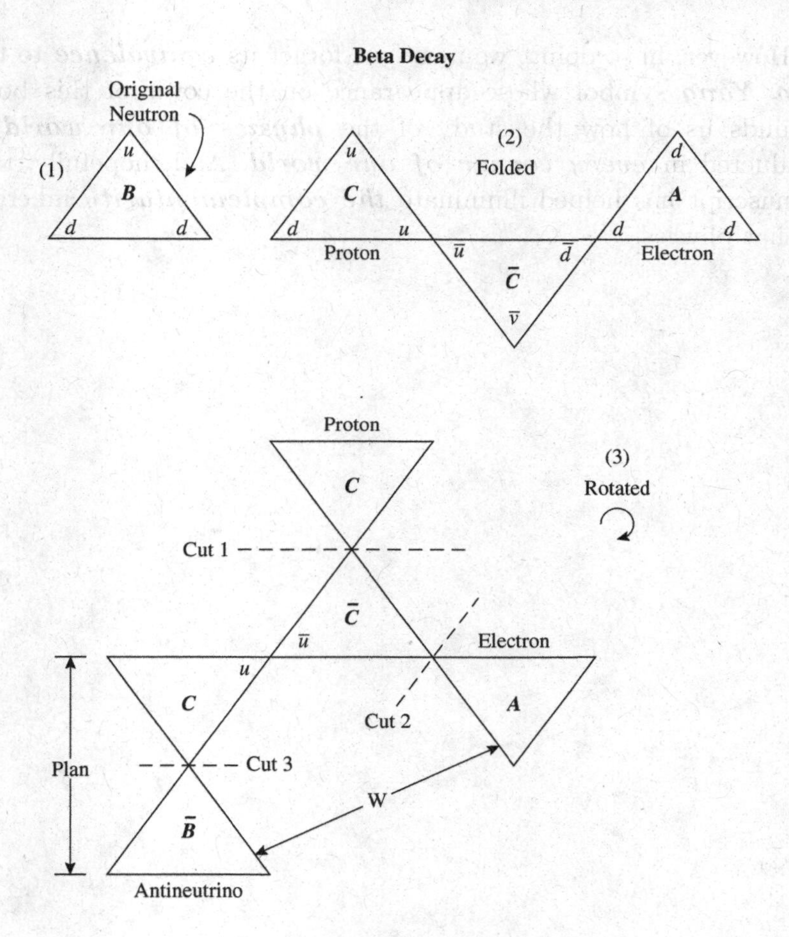

Fig. A.1. Beta Decay

things, highlights the essence of the ***growth of living creatures***! (See Section 5, "the Signature of complementarity: Replication Unleashed".)

So: we have a bit of a bonus here, one that ***Fermi and Yang*** might well be proud of! And, once more, ***Matrix m*** has ***exhibited its importance*** — its <u>ubiquitous</u> ***importance*** — all the way from ***life*** as we know it down to the ***elementary particles of our universe***! Think about it!

However, in so doing, we must not forget its ***equivalence*** to the ***Yin Yang*** symbol whose appearance on the cover of this book reminds us of how the study of the ***physics of our world*** is conducted in ***every corner of our world***. And, hopefully, this manuscript has helped illuminate ***the complementarity*** inherent in that physics.

Au Revoir

Yes indeed: it would be nice to meet once again. Realistically however, it can't occur within the pages of another technical book (not this one!) that I may or will have authored. As you read these words, I must tell you that this is the second, and last that I anticipate having published by the time I have, or will have turned 95 and that's just a matter of a few months. Writing a technical book, I discovered, is a lot of work, especially if one tries to make not too many mistakes (or express downright falsehoods!).

Nevertheless, I'm really glad I had the opportunity to write this book and its predecessor, my BoR as well (Avrin 2015). I believe the subject matter in each case to be important and to feature a unique point of view that ought to be expressed. As you may recall, such a viewpoint begins with the notion that the elementary particles of physics are ***Sine Gordon Solitons*** in ***Spacetime*** and, as noted in the BoR, page 239: "—the system of sine-Gordon Solitons are generally regarded as realizations of QFT (Quantum Field Theory) —".

So: we have at least a minimal connection to ***Quantum theory***. And furthermore, viewing our basic particles as concatenations of torus knots whose inherent ***curvature*** implies a minimal connection to ***General Relativity.***

In other words, our basic particles appear to provide, at least, a minimal connection to <u>**both**</u> the ***quantum world*** and ***General Relativity.*** In that sense, therefore, Quantum Mechanics and General Relativity are thus ***complementarily connected***! There's more but this may be enough to make the point!

However, it's always a good idea to invoke well-known expertise so here's **Albert Einstein's** view on what's fundamental as mentioned in the BoR: In page 196 of that book, the justly famous physicist Professor ***John A. Wheeler*** who taught at Princeton for many years and thus became a good friend of Einstein's quotes him as having "— a vision of a totally geometric world, a world in which everything was composed ultimately only of spacetime" and the BoR goes on to say that Einstein's vision sounds a lot like the ***solitonic*** particles featured in that book.

Which are, of course, the particles featured in this book, the one you've been reading! Perhaps Einstein would have agreed with our claim — that QM and GR are ***complementarily*** connected, that is. Actually, there are more relationships that demonstrate it but this is all I want to talk about now. Au Revoir: Be well and enjoy your life. I hope you enjoyed this book too. And it wouldn't hurt my feelings if you told someone else about it.

Jack Avrin

References

[1] Aharonov Y and Bohm D. 1959. "Significance of electromagnetic potentials in quantum theory". *Physical Review.* **115**: 485–491.

[2] Alexander JW. 1923. A lemma on systems of knotted curves. *Proc. Nat. Acad. Sci. USA* **9**:93.

[3] [APS] American Physical Society. 2001. This Month in Physics History. September 4, 1821 and August 29, 1831: Faraday and electromagnetism. *APS News* **10**(8):2.

[4] [APS] American Physical Society. 2002. This Month in Physics History. October 1900: Planck's formula for black body radiation. *APS News* **11**(9):2.

[5] [APS] American Physical Society. 2006a. This month in physics history. March 20, 1800: Volta describes the electric battery. *APS News* **15**(3):2.

[6] [APS] American Physical Society. 2006b. This month in physics history. May, 1911: Rutherford and the discovery of the atomic nucleus. *APS News* **15**(5):2.

[7] [APS] American Physical Society 2007. This Month in Physics History. November 1887: Michelson and Morley report their failure to detect the luminiferous ether. *APS News* **16**(10):2.

[8] [APS] American Physical Society. 2008. This month in physics history. July 1820: Oersted and electromagnetism. *APS News* **17**(7):2.

[9] [APS] American Physical Society. 2010. This Month in Physics History. July 1849: Fizeau publishes results of speed of light experiment. *APS News* **19**(7):2.

[10] [APS] American Physical Society. 2016a. This month in physics history. June 1785: Coulomb measures the electric force. *APS News* **25**(6):2.

[11] [APS] American Physical Society. 2016b. This month in Physics history. July 1816: Fresnel's evidence for the wave theory of light. *APS News* **25**(17):2.

[12] Avrin JS. 2005. A visualizable representation of the elementary particles. *J. Knot Theory Ramifications* **14**:131–176.

[13] Avrin JS. 2008. Flattened Möbius strips: their physics, geometry and taxonomy. *J. Knot Theory Ramifications* **17**: 835–876.

[14] Avrin JS. 2011. Torus knots embodying curvature and torsion in an otherwise featureless continuum. *J. Knot Theory Ramifications* **20**:1723–1739.

[15] Avrin JS. 2012a. On the taxonomy of flattened Möbius strips. *J. Knot Theory Ramifications* **21**: 1250004.

[16] Avrin JS. 2012b. Knots on a torus: a model of the elementary particles. *Symmetry* **4**:39–115.

[17] Avrin JS. 2015. *Knots, braids and Möbius strips. Particle physics and the geometry of elementarity: an alternative view.* World Scientific.

[18] Barrow JD. 2007. *New theories of everything: the quest for ultimate explanation.* Oxford University Press.

[19] Bartusiak M. 2003. *Einstein's unfinished symphony: listening to the sounds of space-time.* Berkeley Books.

[20] [BBC] British Broadcasting Corporation. 2021. Sir Isaac Newton. Learning English. Moving Words. [Accessed 9 June 2021]. https://www.bbc.co.uk/worldservice/learningenglish/movingwords/shortlist/newton.shtml.

[21] Bohr N. 1913. On the constitution of atoms and molecules. *Philos. Mag.* **26**: 1.

[22] Bozsaky D. 2010. The historical development of thermodynamics. *Acta Technica Jaurinensis* **3**:3.

[23] Byers, N. 1999. E. Noether's discovery of the deep connection between symmetries and conservation laws. In: *Israel Mathematical Society Conference Proceedings. Gelhart Research Institute of Mathematical Sciences and Emmy Noether Research Institute of Mathematics.* Vol. 12. Bar-Ilan University. [Accessed 6-7-2021]. http://arxiv.org/abs/physics/9807044.

[24] Carnot S, Clapeyron E and Clausius R. 1960. Mendoza E (ed.). *Reflections on the motive power of fire – and other papers on the second law of thermodynamics.* Dover Publications.

[25] Clifford WK. 1870. On the space-theory of matter. *Proceedings of the Cambridge Philosophical Society* (1864-1876 (Printed 1876) **2**: 157. Available online at wikisource.org. [Accessed 11 June 2021.] https://en.wikisource.org/wiki/On_the_Space-Theory_of_Matter.

[26] Curtright TL, Fairlie DB and Zachos CK. 2014. *A concise treatise on quantum mechanics in phase space.* Imperial College Press.

[27] Eilbeck C. 1995. John Scott Russell and the solitary wave. Heriot-Watt University. Department of Mathematics. [Accessed 7 June 2021.] https://www.macs.hw.ac.uk/.~chris/scott_russell.html.

[28] Einstein A. 1919. Do gravitational fields play an essential role in the structure of elementary particles of matter? *Preussische Akademie der Wissenschaften Berlin (Mathe matical Physics)* **1919**: 349.
[29] Einstein A. 1946. The meaning of relativity. Methuen.
[30] [ETHW] Engineering and Technology Wiki 2019a. Maxwell's equations. [Accessed 2020 January 15]. https://ethw.org/Maxwell%27s_Equations.
[31] [ETHW] Engineering and Technology Wiki 2019b. Heinrich Hertz. [Accessed 2020 January 15]. https://ethw.org/Heinrich_Hertz.
[32] [ETHW] Engineering and Technology History Wiki. 2021. James Clerk Maxwell. [Accessed 2021 June 6]. https://ethw.org/James_Clerk_Maxwell.
[33] Famous Scientists. 2014. Albert Einstein. famousscientists.org. [Accessed 9 June 2021]. https://www.famousscientists.org/albert-einstein/.
[34] Famous Scientists. 2016. Arthur Compton. famousscientists.org. [Accessed 13 June 2021]. www.famousscientists.org/arthur-compton/.
[35] Famous Scientists. 2021. Joseph-Louis Lagrange. famousscientists.org. [Accessed 9 June 2021]. https://www.famousscientists.org/joseph-louis-lagrange/.
[36] Fermi E. 1937. *Thermodynamics.* Prentice-Hall.
[37] Fermi E and Yang CN. 1949. "Are Mesons elementary particles?" *Phys. Rev.* **76**:1739
[38] Feynman, RP, Morinigo FB and Wagner WG. 1995. *Feynman lectures on gravitation.* Hatfield B editor. Westview Press.
[39] Flapan E (2000). *When topology meets chemistry: a topological look at molecular chirality.* Cambridge University Press.
[40] Fleming J. 2017. Carl-Gustaf Rossby: Theorist, institution builder, bon vivant. *Physics Today* **70**: 50.
[41] Greenstein G and Zajonc AG. 1997. *The quantum challenge: modern research on the foundations of quantum mechanics.* Jones and Bartlett.
[42] Gross D. 1992. In: Davies PCW, Brown J editors. *Superstrings: a theory of everything?* Cambridge University Press. Chapter 6; p. 140.
[43] Hatfield G. 2018. René Descartes. Stanford Encyclopedia of Philosophy (Summer 2018 Edition). Zalta EN editor. [Accessed 11 June 2021]. https://plato.stanford.edu/archives/sum2018/entries/descartes/.
[44] Janssen M and Renn J. 2015. History: Einstein was no lone genius. *Nature* **527**:298.
[45] Kleppner D and Jackiw R. 2000. One hundred years of quantum physics. *Science* **289**:893.
[46] Krider P. 2006. Benjamin Franklin and lightning rods. *Physics Today* **59**:42

[47] Lam KS. 1992. Lam KS. 2009. *Non-relativisitic quantum theory: dynamics, symmetry, and geometry.* World Scientific.

[48] Lanczsos C. 1949. *Variational principles of mechanics.* University of Toronto Press.

[49] Landau LD and Lifshitz EM. 1969. *Mechanics.* 2d ed. Pergamon Press.

[50] LibreTexts. 2021. Phase diagrams. The LibreTexts Project. [Accessed 9 June 2021]. https://chem.libretexts.org/@go/page/1535.

[51] Look BC. 2013. Gottfried Wilhelm Leibniz. Stanford Encyclopedia of Philosophy (Spring 2020 Edition). Zalta EN editor. [Accessed 9 June 2021). https://plato.stanford.edu/archives/spr2020/entries/leibniz/.

[52] Lorentz HA. 1909. *The theory of electrons and its application to the phenomena of light and radiant heat.* Columbia University Press.

[53] Lorentz HA. 1902. The theory of electrons and the propagation of light. Nobel Prize lecture. The Nobel Prize. [Accessed 2021 January 2020]. https://www.nobelprize.org/prizes/physics/1902/lorentz/lecture/.

[54] Lykken J and Spiropulu M. 2014. Supersymmetry and the crisis in physics. *Scientific American* **310**(5):34–9.

[55] Machamer P and Miller DM. 2021. *Galieo Galilei. The Stanford Encyclopedia of Philosophy.* Summer 2021 ed. Zalta EN editor. Forthcoming URL: https://plato.stanford.edu/archives/sum2021/entries/galileo/.

[56] Maldacena J and Susskind L. 2013. Cool horizons for entangled black holes. *Fortschr. Phys.* **61**:781.

[57] Minkowski H. 1908. Raum und zeit. Physikalische Zeitschrift 10: 104–111. Translation available at: Space and Time: Minkowski's Papers on Relativity. The Minkowski Society. [Accessed 2021 January 21]. https://www.minkowskiinstitute.org/mip/MinkowskiFreemiumMIP 2012.pdf.

[58] Moscowitz C. 2017. Tangled up in spacetime. Scientific American, 26 October 2016. [accessed 11-5-2020]. https://www.scientificamerican.com/article/tangled-up-in-spacetime/.

[59] Nadler S. 2020. *Baruch Spinoza. Stanford Encyclopedia of Philosophy* (Summer 2020 Edition). Zalta EN editor. [Accessed 11 July 2021]. https://plato.stanford.edu/archives/sum2020/entries/spinoza/.

[60] Neuenschwander DE. 2011. *Emmy Noether's wonderful theorem.* Johns Hopkins University Press.

[61] Newton I. 1729. *Mathematical principles of natural philosophy.* Motte A translator. London. Available online at wikisource.org. [Accessed 11 June 2021.] https://en.wikisource.org/wiki/The_Mathematical_Principles_of_Natural_Philosophy_(1729)/Definitions#Scholium.

[62] Newton I. 1730. *Optics, or, a treatise of the reflections, refractions, inflections and colours of Light*, 4th ed. London. Excerpted by:

Fordham University. Modern History Sourcebook. Isaac Newton: Optics. [Accessed 8 June 2021.] https://sourcebooks.fordham.edu/mod/newton-optics.asp.

[63] [NIST] National Institute of Standards and Technology. 2018. Ampere: history. [Accessed 2021 June 5]. https://www.nist.gov/si-redefinition/ampere-history.

[64] [NLM] National Library of Medicine. The discovery of the double helix, 1951–1953. National Institutes of Health. [Accessed 11 June 2021]. https://profiles.nlm.nih.gov/spotlight/sc/feature/doublehelix.

[65] Nobel Foundation. 2021. The Nobel prize in physics 1954. [Accessed 68 June 2021]. https://www.nobelprize.org/prizes/physics/1954/summary/.

[66] O'Connor JJ and Robertson EF. 2015. Hermann Minkowski. MacTutor. University of Saint Andrews. [Accessed 9 June 2021]. https://mathshistory.st-andrews.ac.uk/Biographies/Minkowski/.

[67] O'Raifeartaigh L. 1997. *The dawning of gauge theory.* Princeton University Press.

[68] Pais A, Jacob M, Olive DI and Atiyah MF. 1998. *Paul Dirac: The man and his work.* Cambridge University Press.

[69] Peebles PJE. 1992. *Quantum mechanics.* Princeton University Press.

[70] Peliti L. 2011. *Statistical mechanics in a nutshell.* Epstein M, translator. Princeton University Press.

[71] Raatikainen P. 2021. *Gödel's incompleteness theorems.* Stanford Encylcopedia of Philosophy. Spring 2021 edition. Zalta EN editor. [Accessed 8 June 2021]. https://plato.stanford.edu/archives/spr2021/entries/goedel-incompleteness/

[72] Schrödinger E. 1944. *What is life? The physical aspect of the living cell.* Cambridge University Press.

[73] Scientific American. 1990. *The laureates anthology: a collection of articles by Nobel Prize winning authors.* Scientific American, 1 January 1990.

[74] Shannon CE. 1948. A mathematical theory of communication. *Bell System Technical Journal* **27**: 379, 623.

[75] Tatum J. 2020. The Fitzgerald-Lorentz contraction. The LibreTexts project. [Accessed 9 June 2021]. https://phys.libretexts.org/@go/page/8462.

[76] Tegmark M. 2014. *Our mathematical universe : my quest for the ultimate nature of reality.* Alfred A. Knopf.

[77] Tietz T. 2018. Augustin-Jean Fresnel and the wave theory of light. SciHi Blog. [Accessed 12 June 2021.] http://scihi.org/augustin-jean-fresnel-wave-theory-light/.

[78] Tod KP. 2006. Spinors and spin coeffici.ents. Encyclopedia of Mathematical Physics. Academic Press. [Accessed 6 June 2021]. https://www.sciencedirect.com/sdfe/pdf/download/eid/3-s2.0-B0125126662000158/first-page-pdf
[79] Torretti R. 1983. *Relativity and geometry.* Pergamon Press.
[80] Vilenkin A. 2015. The beginning of the universe. Inference 1(4). [Accessed 11 June 2021]. https://inference-review.com/article/the-beginning-of-the-universe.
[81] Watson JD. 1968. *The double helix.* Atheneum Press.
[82] Whittaker ET. 1989. *A treatise on the analytical dynamics of particles and rigid bodies.* 4th ed. Cambridge University Press.
[83] Wilczek F. 2012. Happy birthday, electron. *Scientific American* **306**:24.
[84] Wilkins DR. 2005. William Rowan Hamilton: mathematical genius. PhysicsWorld. [Accessed 9 June 2021]. https://physicsworld.com/a/william-rowan-hamilton-mathematical-genius/.
[85] Witten E. 1992. In: Davies PCW, Brown J editors. *Superstrings: a theory of everything?* Cambridge University Press. Chapter 3; p. 90.
[86] Woodward PM. 1953. *Probability and information theory, with applications to radar.* Pergamon Press.

Vol. 63: *Board Games: Throughout the History and Multidimensional Spaces*
by J. Kyppö

Vol. 62: *The Mereon Matrix: Everything Connected through (K)nothing*
edited by J. B. McNair, L. Dennis & L. H. Kauffman

Vol. 61: *Numeral Systems with Irrational Bases for Mission-Critical Applications*
by A. Stakhov

Vol. 60: *New Horizons for Second-Order Cybernetics*
edited by A. Riegler, K. H. Müller & S. A. Umpleby

Vol. 59: *Geometry, Language and Strategy: The Dynamics of Decision Processes — Vol. 2*
by G. H. Thomas

Vol. 58: *Representing 3-Manifolds by Filling Dehn Surfaces*
by R. Vigara & A. Lozano-Rojo

Vol. 57: *ADEX Theory: How the ADE Coxeter Graphs Unify Mathematics and Physics*
by S.-P. Sirag

Vol. 56: *New Ideas in Low Dimensional Topology*
edited by L. H. Kauffman & V. O. Manturov

Vol. 55: *Knots, Braids and Möbius Strips*
by J. Avrin

Vol. 54: *Scientific Essays in Honor of H Pierre Noyes on the Occasion of His 90th Birthday*
edited by J. C. Amson & L. H. Kauffman

Vol. 53: *Knots and Physics (Fourth Edition)*
by L. H. Kauffman

Vol. 52: *Algebraic Invariants of Links (Second Edition)*
by J. Hillman

Vol. 51: *Virtual Knots: The State of the Art*
by V. O. Manturov & D. P. Ilyutko

Vol. 50: *Topological Library*
Part 3: Spectral Sequences in Topology
edited by S. P. Novikov & I. A. Taimanov

Vol. 49: *Hopf Algebras*
by D. E. Radford

Vol. 48: *An Excursion in Diagrammatic Algebra: Turning a Sphere from Red to Blue*
by J. S. Carter

Vol. 47: *Introduction to the Anisotropic Geometrodynamics*
by S. Siparov

Vol. 46: *Introductory Lectures on Knot Theory*
edited by L. H. Kauffman, S. Lambropoulou, S. Jablan & J. H. Przytycki

Vol. 45: *Orbiting the Moons of Pluto*
Complex Solutions to the Einstein, Maxwell, Schrödinger and Dirac Equations
by E. A. Rauscher & R. L. Amoroso

Vol. 44: *Topological Library*
Part 2: Characteristic Classes and Smooth Structures on Manifolds
edited by S. P. Novikov & I. A. Taimanov

Vol. 43: *The Holographic Anthropic Multiverse*
by R. L. Amoroso & E. A. Ranscher

Vol. 42: *The Origin of Discrete Particles*
by T. Bastin & C. Kilmister

Vol. 41: *Zero to Infinity: The Fountations of Physics*
by P. Rowlands

Vol. 40: *Intelligence of Low Dimensional Topology 2006*
edited by J. Scott Carter et al.

Vol. 39: *Topological Library*
Part 1: Cobordisms and Their Applications
edited by S. P. Novikov & I. A. Taimanov

Vol. 38: *Current Developments in Mathematical Biology*
edited by K. Mahdavi, R. Culshaw & J. Boucher

Vol. 37: *Geometry, Language, and Strategy*
by G. H. Thomas

Vol. 36: *Physical and Numerical Models in Knot Theory*
edited by J. A. Calvo et al.

Vol. 35: *BIOS — A Study of Creation*
by H. Sabelli

Vol. 34: *Woods Hole Mathematics — Perspectives in Mathematics and Physics*
edited by N. Tongring & R. C. Penner

Vol. 33: *Energy of Knots and Conformal Geometry*
by J. O'Hara

Vol. 32: *Algebraic Invariants of Links*
by J. A. Hillman

Vol. 31: *Mindsteps to the Cosmos*
by G. S. Hawkins

Vol. 30: *Symmetry, Ornament and Modularity*
by S. V. Jablan

Vol. 29: *Quantum Invariants — A Study of Knots, 3-Manifolds, and Their Sets*
by T. Ohtsuki

Vol. 28: *Beyond Measure: A Guided Tour Through Nature, Myth, and Number*
by J. Kappraff

Vol. 27: *Bit-String Physics: A Finite and Discrete Approach to Natural Philosophy*
by H. Pierre Noyes; edited by J. C. van den Berg

Vol. 26: *Functorial Knot Theory — Categories of Tangles, Coherence, Categorical Deformations, and Topological Invariants*
by David N. Yetter

Vol. 25: *Connections — The Geometric Bridge between Art and Science (Second Edition)*
by J. Kappraff

Vol. 24: *Knots in HELLAS '98 — Proceedings of the International Conference on Knot Theory and Its Ramifications*
edited by C. McA Gordon, V. F. R. Jones, L. Kauffman, S. Lambropoulou & J. H. Przytycki

Vol. 23: *Diamond: A Paradox Logic (Second Edition)*
by N. S. Hellerstein

Vol. 22: *The Mathematics of Harmony — From Euclid to Contemporary Mathematics and Computer Science*
by A. Stakhov (assisted by S. Olsen)

Vol. 21: *LINKNOT: Knot Theory by Computer*
by S. Jablan & R. Sazdanovic

Vol. 20: *The Mystery of Knots — Computer Programming for Knot Tabulation*
by C. N. Aneziris

Vol. 19: *Ideal Knots*
by A. Stasiak, V. Katritch & L. H. Kauffman

Vol. 18: *The Self-Evolving Cosmos: A Phenomenological Approach to Nature's Unity-in-Diversity*
by S. M. Rosen

Vol. 17: *Hypercomplex Iterations — Distance Estimation and Higher Dimensional Fractals*
by Y. Dang, L. H. Kauffman & D. Sandin

Vol. 16: *Delta — A Paradox Logic*
by N. S. Hellerstein

Vol. 15: *Lectures at KNOTS '96*
by S. Suzuki

Vol. 14: *Diamond — A Paradox Logic*
by N. S. Hellerstein

Vol. 13: *Entropic Spacetime Theory*
by J. Armel

Vol. 12: *Relativistic Reality: A Modern View*
edited by J. D. Edmonds, Jr.

Vol. 11: *History and Science of Knots*
edited by J. C. Turner & P. van de Griend

Vol. 10: *Nonstandard Logics and Nonstandard Metrics in Physics*
by W. M. Honig

Vol. 9: *Combinatorial Physics*
by T. Bastin & C. W. Kilmister

Vol. 8: *Symmetric Bends: How to Join Two Lengths of Cord*
by R. E. Miles

Vol. 7: *Random Knotting and Linking*
edited by K. C. Millett & D. W. Sumners

Vol. 6: *Knots and Applications*
edited by L. H. Kauffman

Vol. 5: *Gems, Computers and Attractors for 3-Manifolds*
by S. Lins

Vol. 4: *Gauge Fields, Knots and Gravity*
by J. Baez & J. P. Muniain

Vol. 3: *Quantum Topology*
edited by L. H. Kauffman & R. A. Baadhio

Vol. 2: *How Surfaces Intersect in Space — An Introduction to Topology (Second Edition)*
by J. S. Carter

Vol. 1: *Knots and Physics (Third Edition)*
by L. H. Kauffman

Printed in the United States
by Baker & Taylor Publisher Services